THE NEW

Gladys Hickman B.A. Ph.D.
*Formerly Senior Lecturer in Education,
University of Bristol, and
Director, Schools Council Geography 14–18
Curriculum Development Project
Sometime University of Ghana and
Makerere University College*

assisted by
Richard Hickman M.A., Dip.T.P.
and Lorna Hickman B.A.

Hodder & Stoughton
LONDON SYDNEY AUCKLAND TORONTO

Acknowledgments

Where credit is due: the authors' appreciation

The preparation of this book would have been impossible without the generous help of individuals and institutions in Africa over the last 40 years. During this period the authors made studies, or worked for longer spells in about 32 of the 50 African countries.

There are several 'generations' of helpers. They were school and university students, farmers, *matutu* drivers, teachers, town planners, writers of research reports on irrigation, soils, fisheries, health (bilharzia), environmental hazards (locusts, mealy bugs), energy conservation and alternative technology. The institutions include helpful government departments, survey offices, libraries, schools, universities and other research institutions, and international agencies such as UNEP in Nairobi and United Nations Regional headquarters in Lagos and Addis Ababa. The management and personnel of inumerable oil, copper, coal, sugar companies and irrigation systems such as the Gezira, gave time to conduct and explain field and factory processes. Equally important are those who pulled us out of mud, sand, or irrigation ditches!

It follows that over the years hundreds of people of great character and ability, some now holding high office in their respective countries, have contributed to this great cooperative enterprise. It would be impossible to name them all here, but they and their contributions (especially in the case studies) are 'alive' and remembered with warmth and gratitude. They will recognise themselves and their work. These people have been our teachers and guides towards the future Africa, '... a harsh but beautiful continent ... made for greatness, freedom and unequalled nobility'.

The author and publisher wish to thank the following for permission to reproduce the illustrations in this book:

Aerofilms Ltd, Figure 12.1 British Telecommunications plc, Figure 16.7 Camera Press Ltd., Figures 5.4, 20.9 J. Allan Cash, Figures 2.12, 9.2, 9.3 John Dominy, Figures 14.2, 23.2 Mark Edwards/Still Pictures, Figure 20.7 FAO, Figures 8.7, 20.2 Firestone, Figure 7.4 Barbara Gunnell, Figure 14.5 John Hillelson Agency, Figure 17.9 (Ernest Cole) Intermediate Technology Development Group, Figure 22.8 IFAD, Figure 19.7 Terence McNally, Figure 17.12 New African, Figure 22.5 Omptima Magazine, Figures 17.2, 21.3 (Stuart Robertson) Oxfam, Figures 19.4, 20.5, 20.6, 23.4 Alan Rake, Figure 8.5 Dr Achim Remde, Figures 1.5, 22.4 RSA Handbook, Figure 17.19 South African Bureau for Information, Figures 17.7, 17.8 Times Newspapers Limited, Figures 2.14, 12.9 Unilever Ltd., Figure 5.2 UAC Timber, Figures 4.5, 7.8 Zambia Information Services, Figure 15.7 Barnaby's Picture Library, Front Cover (Nairobi) The Daily Telegraph, Front Cover (Satellite) John Dominy, Back Cover (Harare townscape) The remaining 23 photographs were taken by the author.

The photograph on the front cover shows part of the central area of Nairobi, Kenya. The satellite photograph shows the weather over Africa on 4 April 1979.

Contents

List of case studies, practical work boxes and map extract boxes — 5
Figure 1.1: Countries of Africa — 6
Figure 1.2: Case studies and land-locked states — 7

PART 1 INTRODUCTION TO AFRICA — 8

1 Studying Africa — 8
The plan of the book — 8
How to get the best from this book — 8
How to make your learning easier — 9
Five major themes — 11

2 The variety of African landscapes — 14
Group 1: The relief and structure of the African continent — 14
Group 2: Climate–vegetation landscapes: forest, desert and savanna — 16
Group 3: How people alter the face of the earth — 20
A classification of African landscapes — 22

3 African countries, African peoples — 25
Africa's past — 25
The colonial 'grab' — 26
The size and population of African countries — 27
Regional groupings — 30

PART 2 REGIONAL STUDIES — 31

Central Africa — 31

4 Gabon: a forested country in equatorial Africa — 31
Changes in Gabon — 31
High, humid closed-canopy forests — 31
How to get country information from summary tables — 36

5 Zaire: an equatorial giant — 37
Rural development and agriculture — 38
The mineral wealth of Zaire — 40
Size, communications, and uneven development — 41

Eastern Africa — 43

6 The plateaus, coastlands and mountains of East Africa — 43
East African climate, weather and land use — 45
People and land in East Africa — 46
The growth of towns — 54

Western Africa — 61

7 The Guinea lands: coast and forest landscapes — 61
Forest landscapes in West Africa — 62

8 Savanna landscapes in West Africa: the Sahel — 72
Rural life in northern Ghana — 72
Contrasts in water use along the River Niger — 75
Drought in the Sahel — 78

9 Urban life and development in West Africa — 82
The agricultural town in West Africa — 82
Lagos: a multi-functional metropolis — 83
Mineral oil in the Niger delta — 86
The Federal Republic of Nigeria: a giant of West Africa — 87
West African cooperation: ECOWAS — 88

Northern Africa — 91

10 The Maghreb — 91
Land-use and relief near Blida, Algeria — 94
Water and climate as resources for development — 95
City life in northern Africa — 95

4 The New Africa

11 The arid lands of northern Africa — 101
   The oasis towns at Touggourt — 102
   The mineral wealth of the desert — 105
   The breakdown of traditional ways of life in Algeria and Mauritania — 106

North-eastern Africa — 107

12 Egypt and the Nile valley — 107
   Agriculture in the Nile valley — 107
   Traditional methods of irrigation — 108
   Increasing agricultural output — 110
   Balancing people and production in present-day Egypt — 111
   The Aswan High Dam — 112
   Town life and industrial development — 114

13 Sudan, Ethiopia, and Somalia — 116
   Cotton in the Gezira — 116
   Common problems in 3 countries — 119
   Drought + war = famine — 120

Southern Africa — 121

14 The front-line states — 121
   The mineral bearing rocks of southern Africa — 122
   The network of railways and export routes — 122
   So what actions are the front-line states taking? — 123

15 Zambia and the copperbelt — 127
   The Zambian copperbelt — 127
   The future for agriculture — 134

16 Zimbabwe: land sharing, farming, and minerals — 137
   Sharing wealth, sharing land — 138
   Minerals in Zimbabwe — 140
   Zimbabwe past and future — 143

South — 145

17 The Republic of South Africa — 145
   The importance of South Africa: an economic and technological giant — 145
   Minerals in South Africa: one hundred years of gold — 146
   Industrial development in South Africa — 150
   Apartheid, a strategy for racial segregation — 153
   Land, food, and people: the agricultural use of the land in South Africa — 156
   Water resources in South Africa — 160
   The future — 161

PART 3 THE CONTINENTAL VIEW — 164

18 African resources and development — 164
   Economic development and wealth — 165
   Resources and development — 167

19 People as a resource — 168
   Counting people — 168
   Disease as a hindrance to happiness, progress, and development in Africa — 170
   Wasted lives, wasted resources — 173

20 Pressure on the land — 175
   How do people feel about the land? — 175
   Rural life in Africa — 175
   Famine and poverty in Africa — 176
   The drive for more food — 178
   The right kind of food — 181

21 Problems of expanding cities — 183
   Contrasts between town and country — 183
   The move to the towns — 183
   Urbanisation: how towns grow — 184
   Unofficial housing: an invaluable form of self-help — 186
   New capital cities — 187

22 The development of industry and infrastructure — 189
   The basic needs for industrial development — 189
   Infrastructure: the back-up to development — 192

23 Planning a continent — 196
   International planning organisations — 196
   Coordination and cooperation, or conflict? — 196

	African cooperation: the need for water development	197
	Africa and the rest of the world: trade and aid	200
	Peace and justice: the chief needs for development	201
Figure 23.5	Areas of opportunity in Africa	204
Figure 23.6	Development projects in Africa	205

Case Studies

1	Logging in Gabon	32
2	Maasai rangelands	46
3	Small independent farms in Uganda	47
4	Mumias sugar estate, Kenya	52
5	Mukono: a duka township	56
6	The study of towns: Mombasa	56
7	Rubber production in Liberia	63
8	Timber and plywood at Sapele, Nigeria	64
9	Farming the oil palm in Zaire and Nigeria	38, 67
10	Cocoa farming in Ghana	68
11	Rural life in Northern Ghana	72
12	Mopti: a river town in Mali	76
13	Kainji dam, Nigeria	77
14	Lagos: a multi-functional metropolis	83
15	Mineral oil in the Niger delta	86
16	Land use and relief near Blida, Algeria	94
17	The oasis towns at Touggourt	102
18	Traditional irrigation in the Nile Valley	108
19	The Aswan High Dam and Lake Nasser	112
20	Cotton in the Gezira	116
21	Copper mining at Nkana-Kitwe, Zambia	127
22	Land tenure contrasts near Makwiro, Zimbabwe	133
23	Coal production at Wankie colliery, Zimbabwe	140
24	Underground gold mining in the Witwatersrand	147
25	Forced removals in South Africa	154
26	Lourensford: pear growing in Cape Province	157
27	Cane sugar production in Natal	158
28	Water projects in South Africa	160
29	Nairobi: the expanding city	185
30	On-site housing improvements in Zambia	187
31	Gaborone: an example of a new capital	188
32	A tourist hotel by Lake Malawi	190

Practical work boxes

1	Climate graphs and work calendars	21
2	Making a line drawing from a textbook photograph	33
3	Making a landscape sketch out of doors	48
4	Practical decision making	55
5	Making your own maps: Mukono as an example	57
6	How to use landscape transects to organise information	93
7	How to answer a resource based examination question	104
8	A Natal sugar estate: interpreting air photographs	159
9	How to draw a simple pie chart (graph)	165
10	How to draw a population pyramid	170

Map extract boxes

1	The port and city of Lagos	96–97
2	An ice-worn landscape: Mount Kenya	98–99
3	A mining town: Nkana-Kitwe, Zambia	130–131
4	Farming contrasts at Makwiro, Zimbabwe	132–133

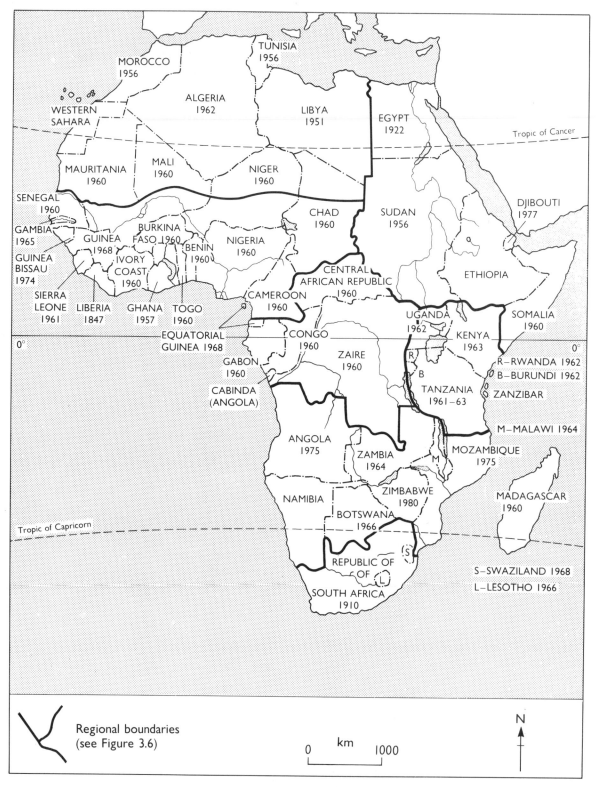

Figure 1.1 Countries of Africa: dates of independence and regional boundaries used in this book

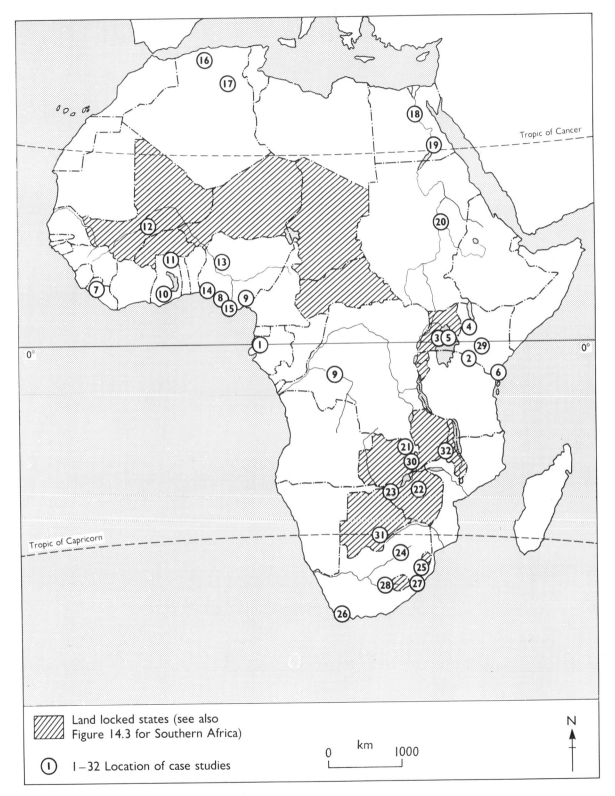

Figure 1.2 Case studies and land-locked states

PART 1 Introduction to Africa

Chapter 1 Studying Africa

The main purpose of this book is to describe conditions and to help students to understand the changes taking place in modern Africa. In this first chapter we are concerned with 3 things:
- How the book is organised so that it *is* possible to begin to understand a great continent – Africa – 'made for greatness, freedom, and unequalled nobility'.
- How the book uses different kinds of study material (resources) to help you to learn more easily. You can use *short cuts* to speed up reading and note making.
- How *you* can make your learning easier and develop study skills. They include those needed for answering examination questions, and for making a living, later on in life.

The plan of the book

There are 3 parts. Parts 1 and 3 have a systematic approach, Part 2 a regional approach.

Part 1

Chapters 1, 2 and 3 are foundation chapters. The studies of landscapes (relief), weather (rainfall, temperature), soils, communications, etc., are systematic studies relevant to the whole continent. They introduce basic situations affecting the development of Africa and can be referred to when studying later parts of the book.

Part 2

Case studies focus on important development themes and what can be learned from them. They are grouped according to the major regions of the continent (see Figure 1.2).

Part 3

This section focuses on the African path to development and selects development situations that are common to much of Africa:
- urbanisation and the drift to towns
- pressure on the land and some solutions
- fuel, energy, and transport problems
- the wise use and conservation of resources of all kinds, from forests to people.

Part 3 shows how people *can* cooperate to find solutions. There are some lessons for other parts of the world.

How to get the best from this book

Finding what you want in a book (retrieval) is usually done:
- by looking at the contents list at the beginning of the book, or
- by looking at the index at the end.

The first gives you a sequence or plan. The second an alphabetical list of the contents. But you may want to find information on:
- commodities
- major cities
- physical geography.

To help you do this quickly and easily there are *signpost tables*, including the *themes table*, on page 12. The signpost symbol ☛ indicates a summary table which organises and classifies a big subject. By showing you what is covered it lets you choose what to study. For example, the table on page 23 lists all the physical geography and shows you where it can be found in the book. The *scale wedge* diagram also appears throughout the book to show how different scales of activity relate to each other.

Different kinds of study resources

This books contains:
1. *Descriptive text* written by the authors.
2. *Study exercises* which are based on photographs, diagrams, or statistics and help you to understand the facts. Study exercises are marked in **bold**. When you work through the study sections you can assemble the answers in your own workbook (jotter) or on the blackboard in class.
3. *Case studies of particular places* which have been chosen because they are typical of many other places or enterprises. The case studies are only useful if some general statements can be developed, that is from one particular industry to industry in general. Sometimes the authors provide the generalisations; sometimes this is left to the students. Examples of generalised 'key' statements are

on page 65 (timber production) and page 86 (urban problems). *Do not read all the studies.* Choose the ones that fit in best with your own area or examination board syllabus.

4 *Practical work boxes.* These boxes interrupt the main text at suitable points to explain especially important methods used by geographers, for example, making simple maps, graphs and field sketches. They help you to carry out work on your own, including field surveys.

5 *Fact boxes.* When an important crop appears for the first time in the book a fact box gives information which can be referred to when studying that crop in other places. Thus the palm products fact box appears on page 38 but can be used for West Africa as well as Zaire. Fact boxes are also used to summarise other information (for example, water development projects, page 199).

6 *Country summaries.* There is a summary table for most of the 48 countries of Africa. It gives trade figures and sometimes a short description.

How to make your learning easier

Do not try to read the whole book

It is important to select parts of the book that relate to your own country, region and examination syllabus, and to do a good deal of *practical* study, that is, collecting local rainfall figures, making field sketches, drawing graphs relevant to your own district. You will understand other places better if you can compare them with your own. Use the contents list, index, and signpost tables to pick out your priority studies.

Speed reading

You can get a rough idea of the content from the *keywords* at the beginning of most chapters. Then use *speed reading*.
- Read the first sentence in each paragraph. It tells you what it is about, that is, its *theme*.
- The rest of the paragraph fills in detail. You only want to read this if you think the detail is important, which it will be for some sections. For other sections it will be less important. Speed reading is also a good way of revising quickly.

Study methods and suggestions for a geographical shorthand

Figure 1.3 shows some examples of different methods that can be used to record information.

1 *Note summaries*, made on the blackboard or in a jotter, can be very brief. They bring out key facts and help you to remember them. Example A on Figure 1.3 is a quick summary of a long study of sugar cane growing and sugar production at Mumias, Kenya on pages 52–54.

2 *Diagram summaries or flow charts* give a simple picture of a complex situation that would be difficult to describe in words. For example, Figure 1.3B and 1.3C are two other ways of showing the stages in growing a crop and preparing it for market.

The word *system* is often used in this book and in world affairs. A dictionary says that 'a system is a set of connected parts'. A diagram of a system makes it easier to see at a glance how the parts fit together and how the world works. Figure 1.3E shows the overlap between two types of farm system, the commercial farm system, and the domestic farm system. Figure 1.3D shows the water cycle (a physical system) in two ways. The sun's energy is the power which starts the transfer of water or water vapour from one part of the system to another. The first diagram shows the detail. In the second diagram only arrows are used to show the connections or flows. There is a similar diagram of arrow flows on page 200.

3 *Line drawings of photographs and labelled landscape sketches.* There are many photographs and diagrams in the book that can be simplified as line drawings like the one in Figure 1.3F. Practical work box 2 on Owendo, page 33, shows you how to do this. Practical work box 3, on page 48, shows you how to simplify and draw a landscape sketch out of doors. Figure 6.7b shows the change of crops grown on different parts of a slope in relation to different soils and the availability of ground water.

4 *Speed drawing.* There is guidance on this in practical work boxes numbers 2 and 3, on pages 33 and 48 and on Figure 1.3F.

5 *Analysis of weather charts.* Use practical work box 1 on page 21 and the weather calendar on page 21, as a starter.

10 Introduction to Africa

A A note summary from Mumias Sugar Estate, P.52

- Position (Figure 6.11 near equator)
- Nucleus estate, 3 400ha (hectares)
- Outgrowers' cane 32 000ha
- Cane cut, transported, milled all year
- Saves Kenya £16 million a year on sugar imports
- Advantages of outgrower system
- Production similar to Tongaat (page 158) but matures in 21 months instead of 24
- Reasons etc

B Flow diagram to show the sequence

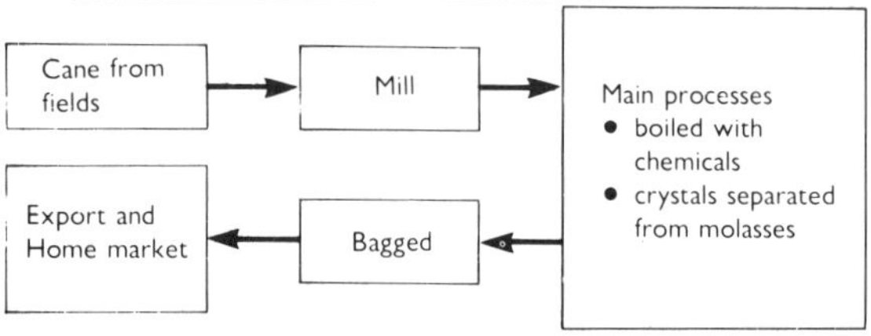

Cane from fields → Mill → Main processes
- boiled with chemicals
- crystals separated from molasses
→ Bagged → Export and Home market

E How two farm systems overlap

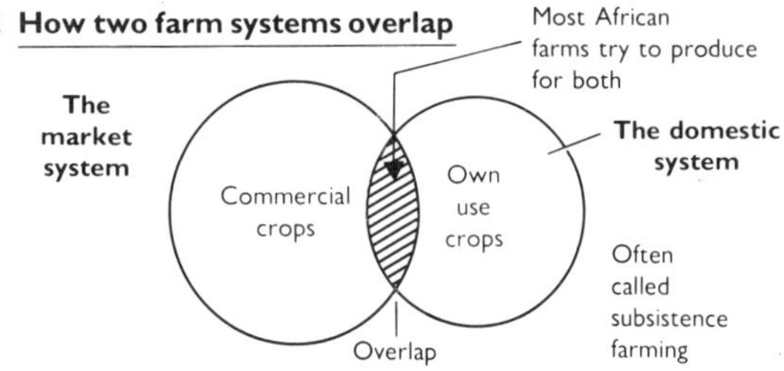

The market system — Commercial crops / Own use crops — The domestic system

Most African farms try to produce for both

Often called subsistence farming

Overlap

C Chart with more detail

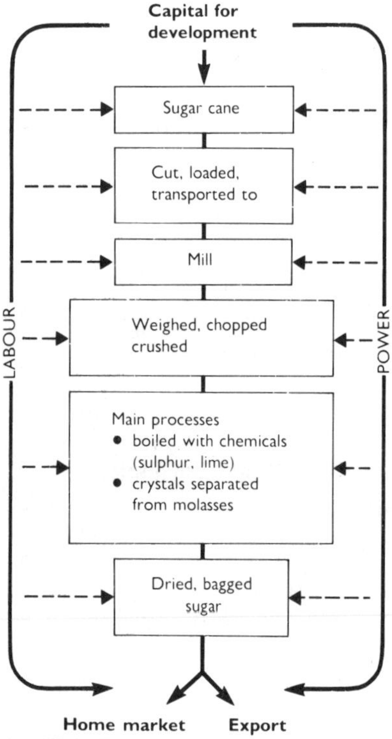

Capital for development ↓

Sugar cane → Cut, loaded, transported to → Mill → Weighed, chopped crushed → Main processes
- boiled with chemicals (sulphur, lime)
- crystals separated from molasses
→ Dried, bagged sugar → Home market / Export

LABOUR / POWER

F Line drawings and labelled sketches

- So you think you can't draw?
- All you need to do is a few lines like this – in your jotter.
- Now check page 99 (from which it was copied) and put some labels on.

D The water cycle

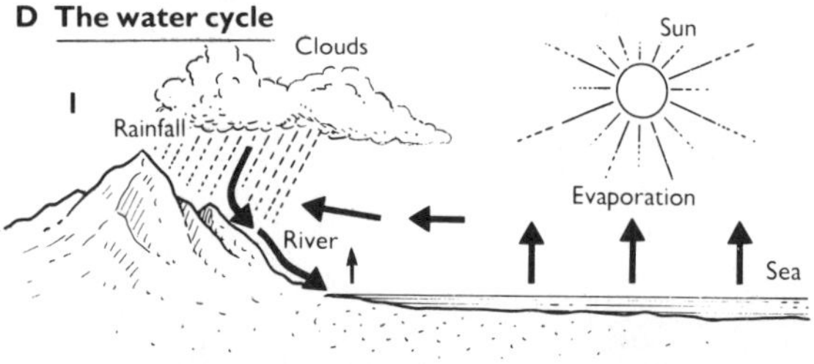

1 Clouds, Rainfall, River, Sun, Evaporation, Sea

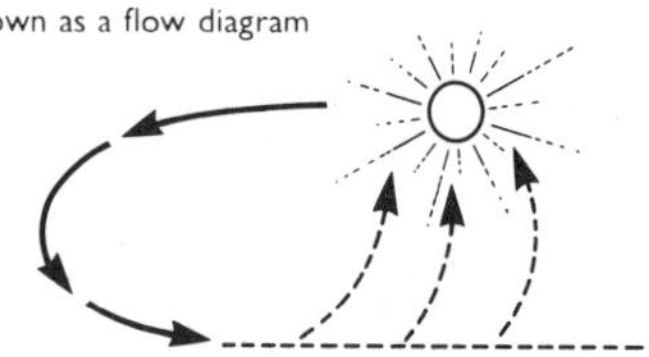

2 Shown as a flow diagram

A similar diagram of arrow-flows could be used to show a trade cycle, a soil fertility cycle; etc.

Figure 1.3 Different ways of showing information

Five major themes

Africa is 8,000 kilometres (km) from Cairo to Cape Town and 7,000 km from Dakar to Mombasa. It contains a vast variety of landscapes and peoples. It is not just a continent of big game safaris, holiday beaches, and reports of famine. There is an everyday Africa where, as in most other parts of the world, there is rapid change.

The problem is to find a way of including enough detail to show the day-to-day lives of ordinary people and yet describe the broad patterns that help us to understand how this great continent works and how it is changing. Five major themes constantly reappear, common threads running through the whole book. These are the themes that really show the unity of Africa.

The *themes table* on page 12 lists the case studies from Part 2 and other studies from Parts 1 and 3 under the 5 main themes and also groups them by region. The themes do *not* appear as chapter headings, and they can only be shown in the table in shorthand. You can gather more examples from the index.

The 5 themes are:
1 The physical environment as a background to development
2 The value of the farming and rural way of life, including its contribution to commercial development
3 Urban trends and industrialisation
4 The place of development projects
5 Development depends on people

Theme 1   The physical environment as the background to development in Africa

The physical landscapes of Africa are varied and magnificent. Along with other features of the environment such as varied climates, soils, water resources, they provide the background against which the people of Africa are developing the continent's resources. At one time people thought that these physical elements decided how land could be used. For most of this century it has been thought that people could overcome the constraints of their environment. Perhaps it is the People's Republic of China that is now on the right track. It suggests that the paths to development are easier when people work '*in harmony* with nature'. People have learned to do this in many parts of Africa.

Theme 2   The value of farming and the rural way of life, including its contribution to commercial development

There is a tendency to think that progress means change, and that new methods are better than old, that industry is better than farming, that towns are more important than villages. All are needed! Key studies show the value of well-tried farm practice, and what can be done to make it easier for people to live well in the countryside.

Examples of this are the different ways of growing cocoa in west Africa; simple terracing to reduce soil erosion; or the value of village kinship organisation for mutual help. Africans are keeping methods that work, and also introducing worthwhile new ideas.

Theme 3   Urban trends and industrialisation

All over the world, towns are increasing in size at the expense of the countryside.

The movement to towns puts a big strain on all urban services. There are not enough homes, so people build 'informal* housing' or shacks. Roads are congested with traffic, drains overflow, money for improvements runs out. This pressure shows in some neighbourhoods of almost every African city, even though the business district is elegantly laid out with many high-rise buildings. Cairo, Harare, Kinshasa or Johannesburg – all have them. Our studies of Lagos or Nairobi partly explain this situation (pages 83 or 185).

Young people especially are attracted by the opportunities and excitement of town life. Towns seem to offer more job opportunities, and chances for advancement. People think they can earn more, that their children will have better schooling, a piped water supply and electricity. But all this costs money.

Theme 4   The place of development projects

Development projects are crucially important. So it is essential to understand how projects have changed over the last 30 years. Many have been only partly successful or have even failed. People's ideas have changed about what really matters.

---

* 'Informal' as used throughout the book means small scale, do-it-yourself.

12 Introduction to Africa

FIVE MAIN THEMES

Location	1. Physical environment as the background to development	2. Farming and rural life (including commercial agriculture)	3. Urban trends and industrialisation	4. Development projects, transport, and trade	5. Development depends on people
CENTRAL AFRICA	Equatorial forests, p. 31 Logging in Gabon, p. 32	Farming the forest, p. 32 Plantation agriculture, p. 39		Transport problems in Zaire, p. 41	How to use trade summaries, p. 34
EASTERN AFRICA	Rift valley, volcanoes, and lakes, p. 16 An ice-worn landscape, p. 99 Mountains as rainmakers, p. 45 Rainfall unreliability, p. 45	Dry zone pastoralists, p. 46 Small farms in Uganda, p. 47 Sugar growing at Mumias, p. 52 Large and small scale agriculture, p. 50	Mukono, a duka township, p. 56 Mombasa, a study of a port, p. 56		Land resettlement in Kenya, p. 51 Rural development through ujamaa in Tanzania, p. 60
WESTERN AFRICA	Soil fertility in the forest zone, p. 66 The Niger flood, p. 75 Drought in the Sahel, p. 78	Rubber production in Liberia, p. 63 Farming the oilpalm bush, p. 67 Cocoa farms in Ghana, p. 68 Rural life in northern Ghana, p. 72	Mopti, p. 76 The growth of towns. Zaria p. 82 Lagos, p. 83, 96 Big city functions, p. 84 Timber production at Sapele, p. 64	Irrigation in the Niger Inland Delta, p. 76 Kainji: a modern multipurpose dam in Nigeria, p. 77 Oil production in the Niger delta, p. 86 ECOWAS, p. 88	Self-help in rural Mali, p. 74
NORTHERN AFRICA	Water and climate as resources for development p. 95 Fossil water p. 102 Aquifers, p. 102	Oasis farming at Touggourt, p. 102	Blida, Algeria, p. 94 City life in northern Africa, p. 95	The mineral wealth of the desert, p. 105	The breakdown of traditional ways of life in Algeria and Mauritania, p. 106
NORTH-EAST AFRICA	The Nile flood, p. 107	Traditional methods of irrigation in the Nile valley, p. 108 Cotton in the Gezira, p. 116	Cairo, Africa's largest city, p. 114	The Aswan High Dam, p. 112 Satellite cities for Cairo, p. 114	Balancing population and production in Egypt, p. 111
SOUTHERN AFRICA	The mineral bearing rocks of southern Africa, p. 122 The geology of the Copperbelt, p. 128	Agricultural progress in Zambia, p. 134 Communal and commercial farming in Zimbabwe, p. 133, 138	Mining and processing copper ore, p. 128 Kitwe: a town on the Copperbelt, p. 130 Wankie coal, p. 140	Railways and export routes for the front-line states, p. 122	Solidarity among the front-line states, p. 123
SOUTH	Gold mining and geology in South Africa, p. 146 Water resources in South Africa, p. 160	Agriculture in South Africa, p. 156	Deep shaft mining on the Rand, p. 147 The PWV industrial triangle, p. 150 Johannesburg, p. 149		Apartheid: a strategy for racial segregation, p. 153 Surplus people, p. 154
THE CONTINENTAL VIEW (PART 3)	The geography of disease and pestilence, p. 170 When the rains fail, p. 176 The expanding desert, p. 177	Rural life in Africa, p. 175 The drive for more food, p. 178	Urbanisation: how towns grow, p. 184 Nairobi: the expanding city, p. 185 New capital cities, p. 187	The basic needs for industry, p. 189 Tourism: an 'invisible' industry, p. 190 Geothermal power, p. 194 Water development projects, p. 197	Too many people? p. 169 Wasted lives and resources, p. 173 Improving informal housing, p. 186 Coordination or conflict? p. 196

Projects are much more than the huge sums of money spent: they must benefit both a country *and* the lives of its people.

In the past projects have played a big part. The dams on the River Nile provided water for the Gezira cotton scheme (page 116). The building of railways made it possible to export coffee from the small shambas of Uganda and groundnuts from Nigeria. Roads, harbours, power sites – all these bring benefits to both countries and people. They provide the basis for further commercial and industrial development. But some people say that they divert attention away from food crops towards commercial and export crop production. Again both are needed.

Now there is a greater appreciation of the value of smaller self-help schemes, working as part of larger projects. They encourage small operators to take part. They include outgrower schemes, cooperative marketing of produce, trading estates like that at Apapa (Lagos) and housing projects like the one in Lusaka. Any country needs both the large multi-purpose, *capital intensive* projects like Kainji or the Tana River schemes in Kenya as well as the smaller *labour intensive* plans bringing modest benefit to many people, such as the horticultural schemes in Kenya or Zimbabwe and the ujamaa villages of Tanzania.

Theme 5 Development depends on people

This theme focuses on the environmental and political hazards which affect people's lives. Diseases such as malaria or bilharzia, and the pests that eat food that humans should have, crucially affect people – their health, well-being and survival. It is not enough to describe them; it is just as important to say what is being done, or can be done to reduce these problems.

Equally, political upheavals and civil wars cause hardship for ordinary people and disrupt production. Even if this is temporary it takes time to get back to normal and get transport and trade going again. This is evident almost everywhere and especially in Mozambique, Sudan, Angola, Ethiopia and Chad. Wars waste people and resources. People can be a country's biggest asset. But they can be a liability if there are too many of them. If the population increases then production must keep in step.

Figure 1.4 Savanna landscape: the dry savanna in the dry season. A lorry on the main road, now tarred, between Gao and Niamey, north of the River Niger in Mali

Figure 1.5 An urban landscape: Kinshasa, Zaire. Broad avenues similar to this were laid out in many African cities in colonial times. The buildings are typical of the Central Business District (CBD)

Chapter 2 The variety of African landscapes

Key words

Natural and man-made landscapes, differences of scale, weather and climate, water balance, groups of landscapes

This chapter gives a simple outline of the physical geography of Africa.
- Where the physical environment relates closely to people's lives or the development of resources, it is included in studies in other parts of the book.
- At the end of the chapter there is a *physical geography checklist*. It allows you to see at a glance where all this information is to be found in the book.

Landscapes are made up of 3 elements:
- Rock structure and relief, that is, the build of the land.
- The vegetation that covers the land surface – a response to the climate (both of these are natural landscapes).
- The use of the land, that is, how people alter the natural landscape. This happens when they farm or irrigate it, build villages, towns, roads and factories, mine it for copper, etc. So there is a great variety of man-made landscapes.

Group 1: The relief and structure of the African continent

Almost the whole of Africa is formed from part of a very old continent which geologists call Gondwanaland. The earth's crust is not fixed. Six huge crust-plates moved gradually allowing parts of former Gondwanaland to drift to their present position in South America, Southern Asia and Australia.

Geologists call the Gondwana rocks in Africa the Basement Complex. The very hard basement rocks do not always show at the surface. Sometimes they are covered by more recent sedimentary rocks, or by huge outpourings from volcanoes, or by the vast extent of Saharan sands, gravels and pebbles.

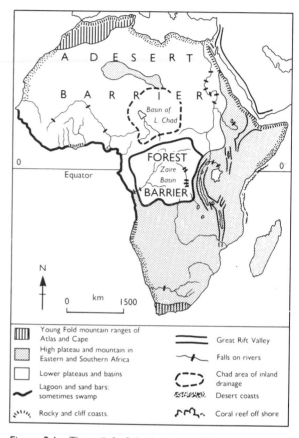

Figure 2.1 The relief of the land and difficult coasts

Use Figure 2.1 and the key to find the following:
1. Two areas of young fold ranges: the Atlas ranges in the north and the Cape ranges in the south.
2. The vast area of plateaus in eastern and southern Africa. There are several levels of upland plains or plateaus, for example at 1,000 metres (m), 1,500 m, and 2,500 m.
3. The lower-level plateaus in northern and western Africa.
4. Low-lying areas forming basins. Name the one that is an area of inland drainage. What does this term mean?
5. List the different types of coast.

The highest land is often volcanic (for example, Mount Kenya, Mount Kilimanjaro, or high land in Ethiopia) or is a huge uptilted part of the crust, for example the Ruwenzori Mountains.

The variety of African landscapes 15

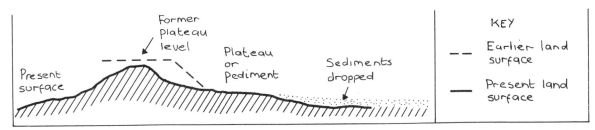

Figure 2.2 Plateau erosion: the land surface is worn back to leave a graded slope (or pediment) at the foot of an escarpment or isolated hill

Find these and the following in your *atlas*:
1 The mountain groups in the Sahara
2 Futa Jallon where the River Niger rises
3 The Jos plateau in northern Nigeria
4 The Drakensberg in the Republic of South Africa

How plateaus are formed

The most widespread relief feature in Africa is plateau land. Figure 2.2 shows how the land surface is worn back to leave a graded slope or pediment at the foot of an upland or isolated hill.

Some special landforms

As well as these large-scale relief features parts of Africa have others that are unique or unusual: *first*, the Great Rift Valley of eastern Africa; *second*, a wide variety of volcanic features; *third*,

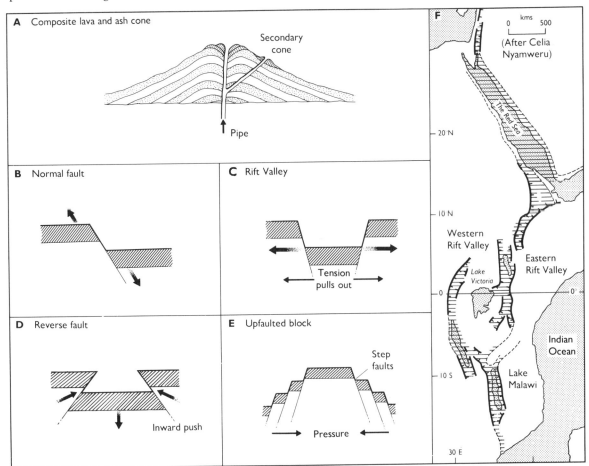

Figure 2.3 Volcanic and rift valley landforms in eastern Africa

16 Introduction to Africa

glaciated (ice-formed) landscapes almost on the equator; and *last*, a great variety of lakes formed in different ways.

Rift Valleys

The Great Rift Valley of eastern Africa forms part of one of the world's best known areas of rifting. It begins 5,000 km to the north of the equator in Israel, where the Sea of Galilee and the Dead Sea occupy part of the Rift Valley. It is often occupied by lakes, some of them very deep, and bordered by sheer cliffs; for example, the floor of Lake Tanganyika is about 600 m below sea level. Figure 2.3 shows some of the Rift Valley landforms.

Volcanic areas

Africa has some of the world's 'hot spots'. There are active and dormant volcanoes (see Figure 2.3A) and electricity is now being generated from steam jets in the Kenyan Rift Valley in East Africa. There are possibilities for geothermal development in other places too (see Chapter 22, page 194). Volcanic areas are productive:
- soils are fertile and varied crops can be grown even inside small craters
- fish can be farmed from crater lakes
- other lakes dry out and concentrate salt, an essential commodity

But there are also hazards. In 1986 villagers near Lake Nyos in the Wum crater area of Cameroon were poisoned by a volcanic gas eruption. Figure 2.4 shows a swarm of explosion craters in west Uganda.

Glaciated landscapes

For information on this type of special landscape see the map extract box on page 99.

Lakes

Both Figure 2.4 and the map extract on page 98, show almost circular lakes. The first type results from volcanic activity, the second from ice action.

There are many other types of lake in Africa.
- The long narrow rift valley lakes occur in down-faulted areas
- Lake Victoria occupies a down-warped area of the East African plateau
- Lake Chad, a lake in an area of inland drainage, is sometimes vast ('MegaChad'), sometimes almost dried up
- There are lagoons and delta lakes along coasts.
- There is a growing number of man-made lakes

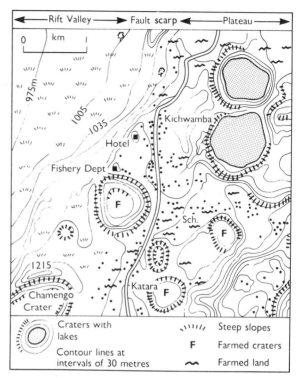

Figure 2.4 A swarm of volcanic craters in west Uganda. Some of the craters are occupied by lakes, others are farmed

such as Lake Nasser, Kariba, Cabora Bassa, and Kainji.

Group 2: Climate–vegetation landscapes: forest, desert and savanna

Weather is what we experience from day to day. Climate is average weather. Figure 2.5 shows climate at different scales.

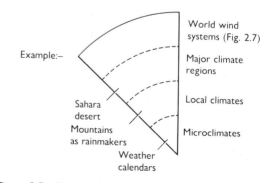

Figure 2.5 Scale wedge: studying climate at different scales

Vegetation landscapes are strongly influenced by climate:
- The equatorial rain forest occurs where the rainfall is abundant, heavy and reliable.
- Arid landscapes – the desert and semi-desert – occur where there is little or no rain and little plant cover.
- When it rains for only half the year, the plants adapt to a long dry season. They are different according to whether rain falls in the cool or the hot season of the year.

Figure 2.6 shows the annual average rainfall of Africa, but everywhere the important factors are not only the amount of rain that falls but also its distribution through the year. It is crucially important to study and understand the pattern of weather and climate in your own and other parts of Africa because the usefulness of the land relates to this.

Where and when it rains in Africa

In theory the winds of the world blow in a very simple pattern, from high pressure to low

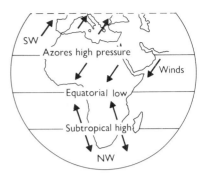

Figure 2.7 World pressure and wind systems

pressure. Figure 2.7 shows how winds over Africa might look, *but*
- The land masses upset the circulation of winds and ocean currents. Winds are drawn in and blow towards the hot continent (low pressure) from cooler, high pressure areas over the oceans in a *convection system*.
- The seasonal shift of the overhead sun and the unequal heating of land and sea move the belts of climate north and south between the tropics of Cancer and Capricorn according to the season.
- Winds are deflected by the earth's rotation.

Check the following points on Figure 2.8 which shows how winds affect weather and climate in Africa.

1. The line of Vs marks the position of the thunder storm zone of convectional rain where air masses coming from different directions meet.
2. The *place* where they meet is a 'heat trough' – the place where the land gets so hot that it abnormally heats the air in contact with it.
3. Such heating of the air causes it to expand and rise and this draws winds inwards. This is a *convergence zone*. It is called the Inter-Tropical Convergence Zone (ITCZ) because it is between the tropics.
4. The storm zone always forms a huge lopsided capital *T*.
5. Its position across the continent controls the weather at different times of the year.
6. Wherever you see it the countries are having their rainy seasons.
7. The high pressure areas are marked H on the wind charts.

These maps can be used for information and reference when you study different parts of the continent. Use both the rainfall and temperature maps, and figures chosen from the climatic statistics on page 24, to describe the sequence of weather through the year.

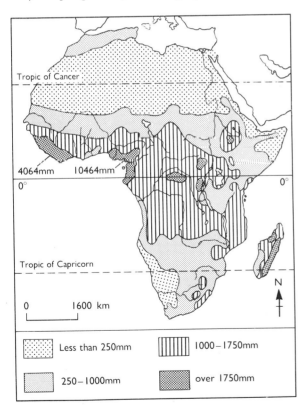

Figure 2.6 Average annual rainfall. This provides only a rough indication of how much rain may fall in a year. It gives no indication of the time of the year when the rain comes, or of reliability. Annual rainfall maps should always be treated with caution

Winds over Africa

It is important to note the extent to which the traditional trade-wind pattern varies in Africa.

- In Figure 2.8A the north-east trades are recognisable but the south-east trades appear only over Madagascar. In the southern Atlantic Ocean the pattern is *completely upset* by the strength of the pull that the hot continent exerts on the air masses.
- A second factor also exerts an influence on the west coast between the equator and the Cape. A vigorous cold current – the Benguela current – wells up and flows northwards along the west coast. This cools the air above it and the land just near the coast, and this situation of cool air blowing to a warmer region is the *reverse* condition from that described on page 17. It is a *subsiding* as opposed to a rising mass of air. Such winds are more likely to absorb moisture than to drop it.
- The contact between warm and cold air causes fogs. At Pointe Noire (the port at the western end of the railway from Brazzaville on the River Zaire) in July the Atlantic breakers were rough and grey. No sun found its way through and the air was cold in contrast to the warmth one expects. Yet Pointe Noire is less than 5° from the equator and the temperature was 10°C instead of 25°C.

Relating rainfall, temperature and evaporation

The *balance* between temperature and rainfall in tropical countries is very important. Where it is very hot there is a high evaporation rate. So a high rainfall total may still not be enough for crops to grow well. A lower rainfall total in a cooler area may be of more use to a farmer.

Figure 2.13A in the practical work box on page 21 shows simple graphs of rainfall and temperature for each month of a year. Figure 2.9 uses temperature and rainfall graphs to show a different kind of information: the *soil moisture* situation, which can mean success or failure to a farmer. Note that monthly rainfall is drawn as 'steps' in Figure 2.13A, but as a line joining points in Figures 2.9A and B.

Now look at Figure 2.9.
1 The weather station in Figure 2.9A is near the equator and 760 m above sea level.

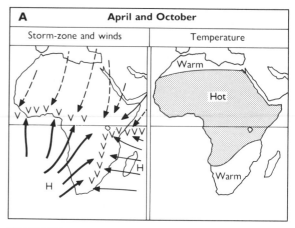

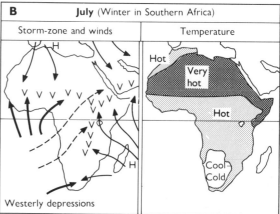

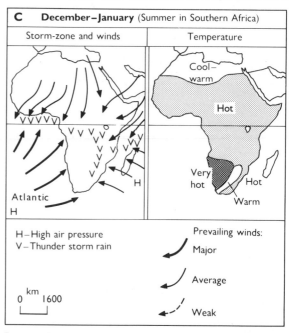

Figure 2.8 Wind systems and climatic seasons: weather charts for Africa

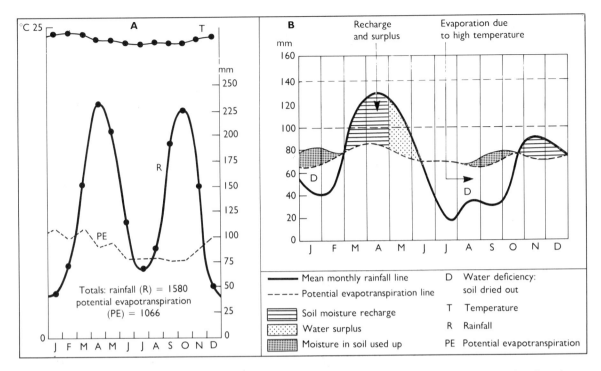

Figure 2.9 'Water balance' diagrams. The balance between rainfall and evaporation is much more complex than these diagrams suggest, but they help you to understand that it is not enough just to learn rainfall totals

2   It has two peaks of heavy rain near the March and September equinox. The total rainfall is 1580 mm.
3   Temperature is very high all year.
4   The high temperature 'dries up' some of the rainfall.
5   The balance of rain, cloud cover, moisture given off by vegetation (transpiration) and drying out by the sun, results in a low *evapotranspiration* level.

The second weather station (Figure 2.9B) is also near the equator, but in East Africa. It has an annual rainfall total of 800 mm. This diagram shows how evapotranspiration affects soil moisture, a situation that is crucial for farmers.
Answer the following questions by checking the symbols in the key.
1   For how many months is the soil at risk from water deficiency (D)?
2   Why is there likely to be a water *deficit*?
3   For how many months after the first dry spell of the year is there rainfall to recharge the soil with water?
4   How long does the water *surplus* last?
5   During which months are crops at risk from the longest dry spell?

How people use climate as a resource

• Our knowledge of the weather and climate of a country depends on records of the weather at hundreds of weather stations.
• Detailed records are kept in order to understand and to *predict* the pattern of the weather.
• But rainfall over much of Africa is *unreliable*.
• Knowing what to expect is an essential part of planning and economic development.

In Africa 20 per cent variability (the percentage by which a year's rain can vary from the average) occurs even in the more humid regions. Sir Joseph Hutchinson said: 'Over much of Africa the number of people who can live in a place is the number of people who can get a drink in the dry season.' All the more reason to store water, conserve run-off in the ground and in tanks and protect soils; and to control rivers.

20 Introduction to Africa

The importance of when it rains

It is difficult for anyone who has not lived in a tropical country to imagine the intensity of a rainstorm (see Figure 2.12). Shortly after, the sun is out again and the ground is steaming. This is 'growing weather' but it is also the time when the most disastrous soil erosion can take place. It matters to people 'when' and 'how' the rain falls, as well as how much comes.

In Navrongo, Northern Ghana there are 5 months that are almost rainless. The ground dries up, the grass withers and only the mango and

Figure 2.12 Storm gutters in the older part of Accra

Figure 2.10 A dry river bed in northern Ghana in December (dry season November–March). Holes have been dug in the gravel bed to find water

Figure 2.11 Pot queue. This stand pipe can only be used for a short time early morning and evening. Families line up their pots to make sure of their water

shea butter trees look green. The water in the rivers dwindles to a few pools (see Figure 2.10). It may take three hours to fill pot with water. Villagers have to walk several kilometres to get water from rivers or wells which still flow. Throughout rural Africa long walks to collect water twice a day take up time that could be used on other work. It is the women and children who walk these kilometres.

The government has helped villages to make earth dams; it has sunk wells and fitted stand pipes with pumps or taps. Look at Figure 2.11 to see what happens.

Group 3: How people alter the face of the earth

When people clear forest, plant crops, build dams, irrigate fields, make terraces or build dykes along a river that is liable to flood, they are adapting the earth's resources to their everyday needs. It is the same process when people build a village or a city, a port, a road. This changes the environment, sometimes for better, sometimes for worse, for example, when soil erosion results

Practical work box 1: climate graphs and work calendars

Warri, Nigeria	Jan	Feb	Mar	Apr	May	June	July	Aug	Sep	Oct	Nov	Dec	
Temperature (°C)	27	27	28	27	27	26	25	26	26	27	27	27	Yearly average 27°C
Rainfall (mm)	33	53	137	229	274	378	391	300	452	323	112	36	Yearly total 2,718 mm

You can make a weather record for your own area using your own figures or from climatic figures in a book.

When a book gives climatic figures for each month of the year the table looks something like the ones for Warri, above.
Figure 2.13A shows these figures as graphs. You can find out what a place is like by asking a few questions.
- What is the total rainfall for the year?
- When does it fall?
- Find the month with the highest rainfall, and the month with the lowest.
- Does this show a clear dry season and wet season?
- If there are two high rainfall seasons we would say that there are two peaks.
- Ask similar questions about temperature.

Figure 2.13B shows how weather is related to yam production in southern Nigeria. It uses 3 symbols to show weather: heavy rainstorms, a mixture of rain and sun, and full sun. You can use these symbols, and others you make for yourself, to record weather for each day, week or month in your own area. The daily ones can 'add up' to give a week's weather and so on. You can add up rainy days, dry days and mixed days.

A simple weather calendar like Figure 2.13B can be used to construct a calendar of farm work through the year for your own area or any part of the world.

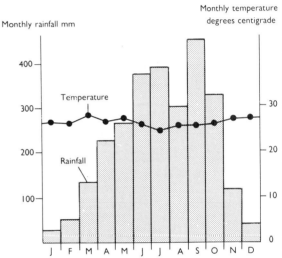

Figure 2.13A Average monthly rainfall and temperature at Warri, Nigeria

	Jan	Feb	Mar	April	May	June	Jul	Aug	Sept	Oct	Nov	Dec	Year
T °C	27	27	28	27	27	26	25	26	26	26	27	27	27
R mm	33	53	137	229	274	378	391	300	452	323	112	36	2718
	☼	☼	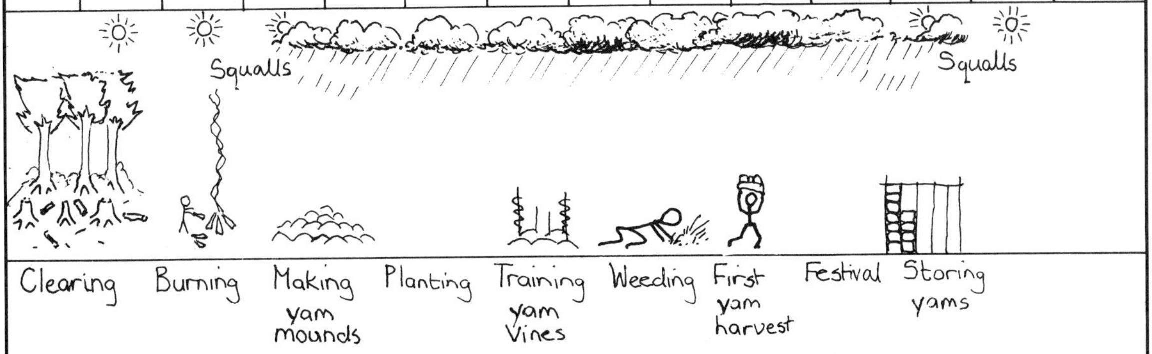 Squalls								Squalls	☼	
	Clearing	Burning	Making yam mounds	Planting		Training yam vines		Weeding		First yam harvest	Festival	Storing yams	

Figure 2.13B A weather and work calendar for Warri

22 Introduction to Africa

Figure 2.14 People **can** improve their own environment

from clearing bush or over-grazing.

In order to think realistically about people and land we must know the problems they encounter in their everyday lives. How well can they earn a living? Why do they live as they do? How do people and governments plan to use the country's resources?

Geographers try to find out why people choose to live in some places and avoid others, that is, why the distribution of population is uneven.

Sometimes this is very simple, for example, if the land is a desert. Sometimes there is no simple answer. For example in parts of Tanzania the vegetation and soils are good and the annual rainfall appears to be sufficient for growing crops; yet few people live there, not nearly as many as the land could support. It is still more puzzling to find that there are cities with hundreds of thousands of people in what is apparently desert; or a railway constructed to a comparatively sparsely peopled area. (The answers come later in the book.)

A classification of African landscapes

We suggested on page 14 that we could classify the landscapes of Africa in 3 different ways. Had we chosen to make a division using only *one* factor, for example Group 2 (vegetation) it would be possible to divide Africa into about 5 'major regional' types, such as equatorial forest, savanna, desert, etc. This would be simple, but it would give an unreal picture of much of Africa.

We have made the more difficult choice of *three* factors because it is nearer to real life and because it helps us to study the interaction between relief, climate–vegetation and people. It is this interaction that gives a place its individual character. In Africa the several extremely varied landscapes are:

1 *Where relief seems to be dominant*
- the coasts
- the plateaus with their rock outcrops
- the highlands and mountains

2 *Where vegetation or its lack is the dominant factor*
- the mangroves of coastal creeks, and other swampy areas
- the forest
- the savanna (wet–dry season lands)
- the desert

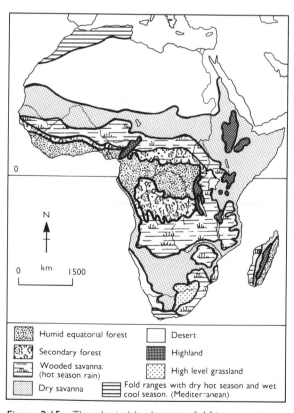

Figure 2.15 The physical landscapes of Africa

3 *Man-made landscapes*
- farmed landscapes of many different kinds
- farmed forest, especially notable because it sometimes still looks like forest
- irrigated landscapes
- mining landscapes
- city landscapes, including the great ports

But it is never as simple as that. For example there are many different kinds of coast. Lagos, Mombasa and Cape Town are all ports but quite different because of their contrasting coastal sites. Nevertheless vast areas are composed of similar types of country. There are 2½ million sq km of forested country, but the forest covers a great variety of landscapes and activities.

The annual rainfall map (Figure 2.6) and the map of physical landscapes (Figure 2.15) are useful for getting to know Africa, provided that we remember that they are too simple. Figure 2.15 is helpful in showing that over extensive sections of Africa there is an essential unity of landscape usually due to climate and vegetation cover, but sometimes due to relief. For example, one can be set down in parts of the wooded savanna country of Northern Nigeria or Zimbabwe (5,000 km apart) and find a plateau with steep domes of weathered granite rising like sudden islands, the scattered trees thorny and bare in the heat of the dry season. But one would know that they are *not* neighbouring places because the people would be a little different.

Ecosystems: the real world

Although we have discussed landscapes where certain factors are dominant, these are not made by either rainfall, temperature, relief, or people *in isolation*. It is the *combination* of all the elements, including people, that gives a place its special character.

The word *ecosystem* is short for *ecological system* and is a convenient shorthand word for a very complex subject. Scientists use it to describe communities in the living world. Ecology comes from a Greek word meaning a home. Thus it is the study of animals and plants in their habitats or environments. Such studies include the ways in which all the parts are related to, and affect each other. They are all *linked*.

An ecosystem includes climate, type of soil, amounts of water, oxygen, food, and light and how these interact to affect the lives of plants, and animals, including people.

Too often writers separate the physical elements from the human. We are all part of the living world, and how we interact and work alters the natural environment. This is the subject of the next chapter.

PHYSICAL GEOGRAPHY CHECKLIST

RELIEF AND LANDFORMS

Mountains and mountain building
Rift valleys and faults, p. 16
Plateaus, p. 15

Landscape features
Volcanoes and craters, p. 16
Geothermal activity, p. 194
Glaciated landscapes, p. 99
Granite tors, p. 138
Coasts, p. 14
Alluvial fan cones, p. 94, 99

Geology and soils
Basement rocks, pp. 14, 122
Mineral extraction:
  oil, p. 86
  copper, pp. 41, 127
  coal, p. 140
  gold, p. 146
Aquifers and water resources, pp. 102, 108
Forest soils, p. 66

Lakes and river basins
Lakes, p. 16
Zaire, p. 37
Niger, p. 75
Nile, p. 107
Zambezi, p. 198

CLIMATE, VEGETATION, AND ECOSYSTEMS

Climate
Weather stations, p. 24
Weather calendars, p. 21
Rainfall-wind systems, p. 17
Rainfall:
  the effect of height, pp. 45
  unreliability, pp. 45, 78, 119
Evapotranspiration, p. 19
Water balance, p. 18

Vegetation
Rain forest, pp. 32, 62
Savanna and Sahel, pp. 62, 72, 119, 120
Desert, pp. 101, 160, 177

Ecosystems and environmental problems
Water cycle, p. 10
Desertification, pp. 101, 106, 177
Pests and diseases, p. 70, 170
Pollution, p. 200

CLIMATIC FIGURES for selected stations

Temperatures are in degrees Centigrade (°C), monthly averages and average for the year.
Rainfall is in millimetres (mm), monthly averages and total for the year

	Jan	Feb	Mar	Apr	May	June	July	Aug	Sep	Oct	Nov	Dec	Year
Algiers, Algeria latitude 36°46'N altitude 59 m													
Temperature	12	13	14	17	19	22	25	26	24	20	16	14	14
Rainfall	112	84	74	41	46	15	0	5	41	79	130	137	764
Touggourt, Algeria latitude 33°07'N altitude 69 m													
Temperature	11	15	18	22	26	31	34	33	30	23	16	12	23
Rainfall	5	10	13	5	5	5	0	0	0	8	13	8	80
Cairo, Egypt latitude 29°52'N altitude 116 m													
Temperature	13	15	18	21	25	28	28	28	26	24	20	15	22
Rainfall	5	5	5	3	3	1	0	0	1	1	3	5	28
Khartoum, Sudan latitude 15°37'N altitude 390 m													
Temperature	25	25	28	32	33	34	32	31	32	32	28	25	30
Rainfall	0	0	0	0	3	8	53	71	18	5	0	0	158
Addis Ababa, Ethiopia latitude 9°20'N altitude 2,450 m													
Temperature	15	17	17	18	18	17	16	16	16	16	14	14	16
Rainfall	13	38	66	86	86	137	279	300	190	20	15	5	1,235
Navrongo, Ghana latitude 10°50'N altitude 200 m													
Temperature	28	30	32	32	31	28	27	26	26	28	28	27	29
Rainfall	0	0	15	51	112	145	203	264	229	69	0	0	1,088
Lagos, Nigeria latitude 6°12'N altitude 3 m													
Temperature	27	28	28	28	28	26	26	26	26	26	28	28	27
Rainfall	28	46	102	150	269	460	279	64	140	206	69	25	1,838
Libreville, Gabon latitude 0°23'N altitude 35 m													
Temperature	30	31	31	31	30	29	28	28	29	29	29	30	30
Rainfall	250	250	325	300	213	25	25	25	100	275	380	200	2,368
Nairobi, Kenya latitude 1°16'S altitude 1,820 m													
Temperature	20	21	21	21	20	18	17	18	19	20	20	20	20
Rainfall	38	53	135	196	132	14	15	25	23	48	102	63	871
Mombasa, Kenya latitude 4°03'S altitude 16 m													
Temperature	28	29	29	28	26	26	25	25	25	26	27	28	27
Rainfall	25	15	61	198	323	107	89	69	64	86	94	61	1,191
Songea, Tanzania latitude 10°41'S altitude 1,153 m													
Temperature	22	23	22	21	19	17	17	18	21	22	23	23	21
Rainfall	274	231	259	112	15	3	0	3	3	10	48	178	1,130
Ndola, Zambia latitude 12°59'S altitude 1,269 m													
Temperature	22	22	22	21	18	15	15	17	22	23	23	22	20
Rainfall	307	235	190	36	4	0	0	1	1	20	127	250	1,171
Harare, Zimbabwe latitude 17°50'S altitude 1,473 m													
Temperature	20	20	19	18	16	13	14	16	19	22	21	20	18
Rainfall	187	173	101	34	11	4	1	2	7	30	97	182	829
Johannesburg, Republic of South Africa latitude 26°14'S altitude 1,665 m													
Temperature	20	20	19	16	13	11	11	13	16	19	19	20	16
Rainfall	114	109	89	38	25	8	8	8	23	56	107	124	709
Durban, Republic of South Africa latitude 29°50'S altitude 5 m													
Temperature	25	26	24	22	20	17	17	18	19	21	23	24	21
Rainfall	112	125	135	85	50	25	25	37	75	125	125	125	1,044
Cape Town, Republic of South Africa latitude 33°54'S altitude 17 m													
Temperature	21	21	20	17	14	13	12	13	14	16	18	19	17
Rainfall	15	8	18	48	79	84	89	66	43	30	18	10	508

Chapter 3 African Countries, African Peoples

Key words

Divided nations, colonial legacy, Islam, average population density, population clusters, areas of opportunity

This chapter divides into 4 parts:
• People in different parts of the continent and their movements.
• How the present country boundaries were formed.
• The size and population of African countries and their population densities.
• The regional groups used in later parts of the book.

Africa's past

Africa is sometimes called 'the cradle of mankind'. This is not just because some of our earliest ancestors evolved in Africa. Some of the earliest civilisations — those of Egypt, Kush, Nubia and Ethiopia — developed in the Nile valley. You can find a *brief summary* of some of the main facts of Africa's history in the fact box on page 26. Consult a book on the history of Africa for more information.

The great historic empires of Africa were little known to Westerners. Look back at Figure 2.1 to see why Africa seemed a 'barrier continent' to people outside.

Figure 3.1 summarises some of the ways outsiders got to know Africa and how people moved into the part of the continent which they now occupy.

Use Figure 3.1, to check the following statements.
1. The area shaded with diagonal lines shows the part of Africa most influenced by Islam. Name the centre from which Islamic culture moved into Africa.
2. The Bantu speaking peoples spread their influence from West Africa down to the south of the continent.
3. The seasonal reversal of the monsoon winds on the east coast made it possible to sail to

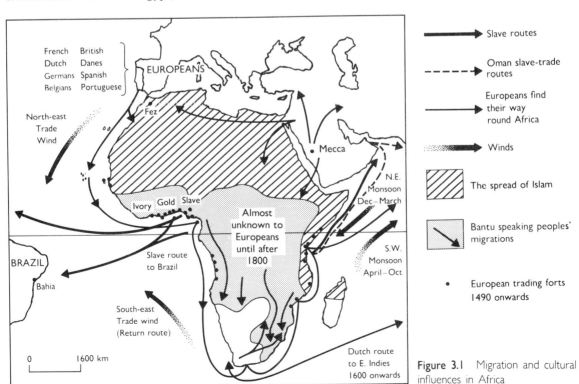

Figure 3.1 Migration and cultural influences in Africa

Oman, Persia, India, South-east Asia and even China. Later, the Portuguese, British and the Dutch used the route to and from South-east Asia and the East Indies.

4 There were 3 main slave routes:
- from the east coast to Oman, Arabia, India
- from the west coast to Brazil, the West Indies and North America
- to South Africa from the East Indies

There have been two periods of 'takeover' in Africa by outsiders: first, the Islamic *cultural* takeover of the northern and eastern part of the continent from the seventh century onwards; and second, the European *political and economic* takeover mainly in the nineteenth century and early twentieth century.

A Kenyan professor, Ali Mazrui, says that no legacy from the colonial past poses more problems than the boundaries of Africa's countries.

The colonial 'grab'

- The European 'scramble for Africa' was confirmed at a conference in Berlin in 1884.
- Nobody asked the African inhabitants what they wanted, and people spoke as though nothing had existed before a European had discovered it.
- Straight lines were drawn on maps with little knowledge of the African reality.
- Seven European countries agreed among themselves to set up 'spheres of influence' extending inland from coastal 'possessions'. Thus the short stretches of coast held by France, Germany, Britain, etc., led to the long narrow countries of West Africa.
- There were some curious boundaries, for example the 'panhandle' of the then Belgian Congo. Both Britain and Belgium wanted the rich copper deposits and agreed to divide the area.

The carving up of the African continent at that time – King Leopold of Belgium described it as 'that magnificent African cake' (see pie graphs, page 165) – has resulted in the present boundaries, and some of the disputes. Many lines, drawn in ignorance, cut through African nations (ethnic groups) leaving them under different colonial powers then, often with different 'official' languages and in different countries today. Sometimes the colonies included peoples who were traditional enemies and who now have the same nationality. Figure 3.2 shows some of the results. Only a few cases are named and not every example is shown.

Facts: African history

- North Africa was periodically invaded over 2,000 years ago by the Greeks, the Romans and the Persians.
- The first far-reaching penetration was the Islamic cultural invasion following the death of Mohammed in 632 AD. Islam spread through the northern half of the continent almost as far as the forests of west Africa, and down the east coast as far south as the island of Madagascar.
- In the sixteenth century the Moroccan and Turkish empires extended across the north of Africa.
- Arab ports flourished on the east coast trading slaves and other goods and the Swahili language developed.
- Bantu speaking people migrated southwards from west Africa to central and southern Africa.
- Many African Kingdoms developed, for example:
  - from the eleventh century onwards – Ghana, Mali
  - from the sixteenth century – the Hausa, Kongo, Mutapa (Zimbabwe)
  - from the eighteenth century – the Ashanti.
- When the European maritime nations began their exploration of the west coast of Africa, on the sea route to the East Indies, they set up trading forts – 'toe-holds' from which to venture into the interior of what they called the Dark Continent.
- Africa was pillaged for slaves during the eighteenth century when the development of plantations (sugar, cotton) in the newly settled Americas triggered the need for abundant labour.
- The second great invasion of Africa was the nineteenth-century 'scramble' when the maritime nations of Europe took over land to form new colonies. The discovery of diamonds and then gold in southern Africa started the rush to take possession of these mineral-rich lands.

During the nineteenth century and the first half of this century almost all the countries were administered by European powers. There were only two independent sovereign states, Ethiopia

African Countries, African Peoples 27

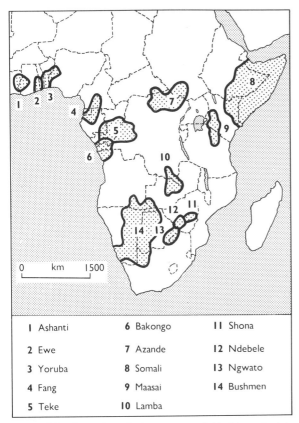

Figure 3.2 Examples of ethnic groups divided by colonial boundaries

1 Ashanti	6 Bakongo	11 Shona
2 Ewe	7 Azande	12 Ndebele
3 Yoruba	8 Somali	13 Ngwato
4 Fang	9 Maasai	14 Bushmen
5 Teke	10 Lamba	

and Liberia. All the countries of Africa except Namibia are now independent, but many of their current problems are the result of the fact that most of their economies are still dependent upon the activities, markets and money supplies developed during the colonial period. A successful colony does not automatically make a united independent state with an economy that works.

The size and population of African countries

There is a list on page 29 of the countries of Africa, their size, and the numbers of people who live there.

Population maps of the world in atlases show great contrasts between the densely settled and the thinly settled parts. In spite of having over 500 million people, Africa falls within the thinly peopled area: the average is about 19 per sq km for the whole continent, compared with nearly 100 per sq km for the continent of Europe. Yet there are parts of Africa where the density increases to 100 persons to a sq km and even up to 1,000 in the delta of the River Nile. In Figure 3.3 the countries of Africa are shaded according to their *average* population density. It shows simple but important contrasts.

1. Africa has three populous zones; north, centre and south.
2. Each is separated from the other by thinly peopled dry zones, the Sahara desert in the north and the Kalahari in the south.
3. West Africa is notable for the number of people who live there. Write down the names of the countries with averages of over 30 people per sq km. What do you notice about this *group* of countries?
4. Rwanda and Burundi are the most densely peopled countries in Africa by average density. They are located north of Lake Tanganyika and east of Lake Kivu. Study the atlas and find out how high the land is. Do the same for Malawi. The section on East Africa provides a number of reasons for these exceptionally high densities.

Such a map as Figure 3.3 has its uses, but can also be misleading. First look at Egypt. The average density map suggests that people are spread evenly all over it, even in the desert. Most people live by the river Nile. Similarly, in north-west Africa, Algeria has the low average density of 9 persons per sq km because of the huge desert

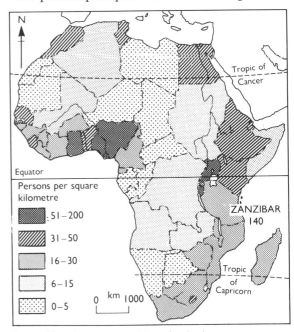

Figure 3.3 Average population density by country

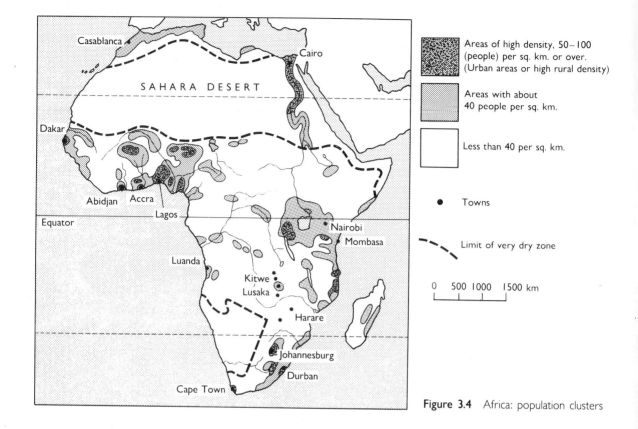

Figure 3.4 Africa: population clusters

area. Yet Algeria has 3 times as many people as Tunisia, which has a density of 40 persons per sq km and a much smaller desert area.

The second population map (Figure 3.4) tries to show where people *really* live.

Study the map and answer the questions.
1. Where is there a great concentration of people in what appears to be *desert*?
2. In which different parts of Africa are the most important clusters of settlement on the coast?
3. Use an atlas to find the *average altitude* (height above sea level) in those parts of East Africa where most people live.
4. Is one of the most extensive of the densely peopled areas the valley of the great River Zaire (like the densely peopled river valleys of India and China) or on the mountain rim further east which is 1,200–2,000 m above sea level? (Refer to Chapter 6)

Because of its great size (Africa is 8,000 km from north to south), the well populated areas are scattered and separated from each other, often by two thousand, or even three thousand kilometres of territory where there is only a thin scattering of people.

One might think of these well-populated areas of Africa as *population oases* in a deserted area – or as *population clusters*. In areas such as these the countryside is well settled and the largest cities of Africa occur, sometimes in almost complete isolation, as in the Copperbelt towns of Zambia. There are four types of population cluster:
- Those near coasts, where some of the cities are ports.
- Certain highland areas of tropical Africa where reliable rainfall, and good soils encourage farming and settlement.
- Economic islands, for example where great mineral wealth or an irrigation or power project encourages development.
- Areas where a combination of factors promotes higher population densities, as in west Africa, the Nile valley, or north-west Africa (the Maghreb).

Figure 3.5 African countries: area and population figures

Country (Date of independence)	Area (000s sq km)	Population (millions, 1986)	People per sq km	Urban population %	Gross national product (1983) Total $ million	Gross national product (1983) Average $ per head
Central Africa						
1. Gabon (1960)	268	1.2	4		2,950	4,250
2. Congo (1960)	342	1.8	5		2,180	1,230
3. Central African Republic (1960)	623	2.7	4	35	690	280
4. Zaire (1960)	2,345	30.9	13		5,050	160
5. Equatorial Guinea	28	0.4	14			
Eastern Africa						
6. Kenya (1963)	580	21.2	36	16	6,450	340
7. Tanzania (1961)	945	22.5	24	17	4,880	240
8. Uganda (1962)	236	16.0	68		3,090	220
9. Rwanda (1962)	26	6.3	238	5	1,540	270
10. Burundi (1962)	28	4.9	175	7	1,050	240
Western Africa						
11. Guinea (1968)	246	6.2	25		1,740	300
12. Guinea Bissau (1974)	36	0.9	25		150	180
13. Sierra Leone (1961)	72	3.7	51		1,230	380
14. Liberia (1847)	111	2.2	20		990	470
15. Ivory Coast (1960)	322	10.2	32	32	6,730	720
16. Togo (1960)	57	3.1	54		790	280
17. Benin (1960)	113	4.0	36	40	1,110	290
18. Cameroon (1960)	475	10.4	22	28	7,640	800
19. Senegal (1960)	197	6.6	34	34	2,730	440
20. Gambia (1965)	11	0.7	58	18	200	290
21. Mali (1960)	1,240	8.4	7	18	1,110	150
22. Burkina Faso (1960)	274	6.8	25	6	1,210	180
23. Niger (1960)	1,267	6.3	5		1,460	240
24. Chad (1960)	1,284	5.1	4	18	400	80
25. Ghana (1957)	239	14.0	59	31	3,980	320
26. Nigeria (1960)	924	98.5	107		71,030	760
Northern Africa						
27. Mauritania (1960)	1,026	1.9	2	23	720	440
28. Morocco* (1956)	713	22.6	32	43	15,620	750
29. Algeria (1962)	2,382	22.4	9		49,450	2,400
30. Tunisia (1956)	164	7.2	44	50	8,860	1,290
31. Libya (1951)	1,760	3.7	2	76	25,100	7,500
North-eastern Africa						
32. Egypt (1922)	1,001	49.6	50	44	31,880	700
33. Sudan (1956)	2,506	22.2	9	20	8,420	400
34. Ethiopia (always independent)	1,222	44.9	37	11	4,860	140
35. Somalia (1960)	638	4.8	7		1,140	250
36. Djibouti	23	0.5	20			
Southern Africa						
37. Angola (1975)	1,247	9.0	7		3,320	470
38. Mozambique (1975)	802	14.3	18	13	2,810	270
39. Namibia (not independent)	824	1.6	2		1,920	1,760
40. Botswana (1966)	582	1.1	2	20	920	920
41. Malawi (1964)	118	7.3	61	12	1,390	210
42. Madagascar (1960)	587	10.3	18	16	2,730	290
43. Zambia (1964)	753	6.9	9	40	3,630	580
44. Zimbabwe (1980)	391	8.4	22	24	5,820	740
South						
45. Republic of South Africa (1910)	1,221	33.2	27		76,890	2,450
46. Lesotho (1966)	30	1.6	51		670	470
47. Swaziland (1968)	17	0.7	39	15	610	890
Totals/averages	30,296 sq km	569.2	19/sq km		$377,770	$714

*Including Western Sahara

Sources: UN Demographic Yearbook and World Bank

Men and women, not just 'population'

It is important to think of people not merely in terms of statistics, but as individuals and groups. We want to know:
- where they live and how they live
- how long they may expect to live
- whether they are racially one group or several
- the rate of population increase.

Racial harmony and the expectancy of life and death are very real problems. Equally important to most people, especially in rapidly developing countries, are the questions 'Can I find a job? Can I get enough food to feed myself and my children? Can I earn enough to pay for their education?

As we study the different parts of the continent we shall find out why some places prosper. These are the areas of opportunity in Africa (page 204). Increasingly, in the modern world, people and nations seek to better themselves and their children; hence each one of the areas of opportunity acts as a magnet. It tends to draw more people, more industries and more invested money to it, and becomes still more populous and important.

Regional groupings

It is convenient to group parts of a large continent together for purposes of study.

Because, on the whole, people want to know about *countries* and because the links between contrasted areas are so important, we have made only some of the studies in the different sections typical of major landscape types.

Check the names of the countries in the 7 main groups shown on the map (Figure 3.6) with the help of an atlas.
1. Central Africa – the Zaire basin, Gabon, the Congo and the Central African Republic.

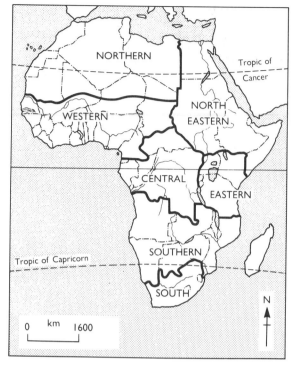

Figure 3.6 Regional groupings in Africa

2. Eastern Africa – the plateau and mountain lands
3. Western Africa – the countries between the Guinea coast and the Sahara Desert, including the Sahel
4. Northern Africa – the Maghreb, and the Sahara including Libya
5. North-eastern Africa – the dry lands of Egypt, the Sudan and the Red Sea coastlands including the valley of the River Nile, Ethiopia and the Horn of Africa
6. Southern Africa – the Front-line States
7. South – The Republic of South Africa, Lesotho and Swaziland

PART 2 Regional studies

Central Africa

Introduction: the selection of key studies

It is impossible to study a whole continent in detail. Instead we have chosen studies which represent some of the most important aspects of life and work in Africa. Each is a 'real life' study which brings out the main points. It is the kind of study you might make in your home area or of an industry in your own country.

A key study can 'stand in' for other similar places. From it you can generalise about a much larger area (for example from local farming to larger commercial farming in similar areas) or from local industries or markets to other places and parts of the continent. It effectively combines the small and the larger scale.

Chapter 4 Gabon: a forested country in equatorial Africa

Key words

Renewable resources, diversified economy, visible trade balance, wealth

Africa is about 3,700 km across from west to east near the equator. It would be natural to think of this area as equatorial, heavily forested or swampy. But in fact there are many variations within these equatorial lands. Not only are there climatic and relief contrasts, there are contrasts in land use and in rural and city living. Libreville, capital of Gabon, Kisangani (Stanleyville) on the River Zaire, Entebbe and Kampala in Uganda, and Nairobi, the metropolis of Kenya, are only a few kilometres distant from this equatorial line.

But the major contrast is between the low-level Zaire Basin on the west and the high-level plateau in East Africa. Chapters 4 and 5, called Central Africa, cover the west and Chapter 6 covers eastern Africa.

Changes in Gabon

The country of Gabon, not much larger than the United Kingdom, is now one of the richest countries south of the Sahara. This is partly because this small country is Africa's second largest oil producer. Petroleum forms 81 per cent of the exports followed by wood products and timber (8 per cent) and the mineral manganese (5 per cent). In 1982 agricultural products were 3 per cent of exports.

- Most oil and natural gas is produced near the coast or from off-shore fields and processed at the Port Gentil refinery. Wealth in oil allows a country to develop chemical and fertiliser industries. But 85 per cent of the crude oil is exported to France, Ivory Coast, Senegal and Curacao.
- Gabon is the fourth world producer of manganese, some going to USSR, India, and South Africa. Uranium is also important and goes to the European Community. There are deposits of iron ore inland.
- You would expect the varied products and exports to benefit the country. They do, but as 80 per cent of the ordinary people are still engaged in agriculture, the real test of Gabon's wealth is how it benefits the farming communities.
- The most recent development in Gabon is the completion of stage 2 of the Trans-Gabon railway to Franceville, the centre for mining manganese. Stage 3 is proposed from Booué to Belinga where the iron ore deposits are.

High, humid, closed-canopy forests

The land area of Africa is 30 million sq km. Originally forest covered nearly *5 million* sq km of country in Africa. Now the total of all types

32 Central Africa

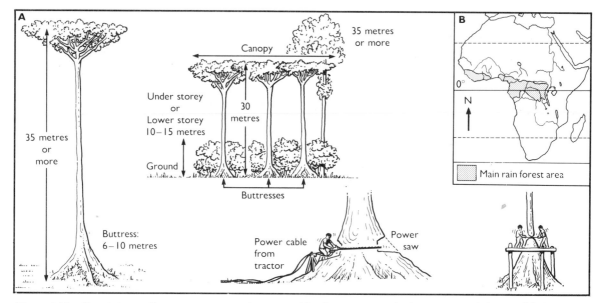

Figure 4.1A Forest terms. Power saws are used at ground level to cut away the buttresses and then to fell the trees. Previously, loggers with hand saws worked from a platform to avoid the buttresses   **Figure 4.1B** Rainforest in Africa

of forest is less than half that (see Figure 4.1B). The high, unaltered, closed-canopy forest – the true *rain forest* – lies away from the coast and main river transport routes. The secondary forest (previously cut, now regrown) is nearer the coast and settlements or other accessible areas. Eighty-five per cent of Gabon is covered by rain forest.

The forest is extremely varied and not all varieties are evergreen, for some trees shed their leaves during the 'dry' season.

The forest is multi-storeyed (Figure 4.1A). The distinctive tall trees are buttressed at the roots, and at the top the crowns of neighbouring trees touch each other and form a 'canopy' about 30 m or more above the ground. From above, this type of forest *looks* impenetrable. The canopy is fairly open, light penetrates near rivers so there is more undergrowth.

Logging in Gabon

Gabon is fortunate in having the world's best supplies of the okoumé tree. Okoumé is a typical forest giant, with a straight smooth trunk branching only at the top. It is a soft wood which has the quality of 'peeling' easily so that it is one of the best woods in the world for making plywood. The absence of branches is important because a branch causes a 'knot' in the wood which is repeated in every layer as it is 'unrolled'.

Farming the forest: forest conservation

The process of felling timber in the depths of the forest, cutting it into sections for hauling to water, floating the timber through rivers and creeks to factory or port for export, is the same in hundreds of places throughout the forest zone of Africa that cut wood for commercial purposes.

In Gabon, three-quarters of the country's timber is felled by large companies in large concessions some distance inland. These companies can afford the heavy equipment required for clearing logging roads and moving the logs. But about 12 per cent is cut by African families owning cutting permits on smaller concessions nearer the coast.

Recently Gabon re-planted 3,000 hectares with okoumé, but it will be 5 years before thinning will give small timber, and 50 years before really important timber is produced.

What has been written about the forest in Gabon might also have been written about other timber exporting countries such as Nigeria, Ghana and Cameroon. All these countries, mainly exporters of logs, could in future use more wood at home, providing work for local people. As well as the very large factories for sawn timber and plywood, the future may see the production of newsprint, packing materials, and cellulose products. If there are more workers earning wages, they can buy more, and this encourages more local industry and services, such as taxis,

Gabon: a forested country in equatorial Africa 33

Practical work box 2: making a line drawing from a textbook photograph

In class
- The purpose of this exercise is to give classroom practice in drawing a landscape sketch and to pinpoint the most important features.
- It is one way to analyse what a picture shows. The method is useful if an examination question includes work on photographs.

Out of doors
It will be much easier to make a landscape sketch out of doors if you have first tried this method out in class.

Method
Look at the photograph, Figure 4.3.
1. Decide on the symbols you want to use for example ~~~ for trees ⌒⌒⌒ for logs. Make up others.
2. Decide how big you want your copy to be.
3. Work out the proportions by dividing the picture into halves and quarters (see sketch). Then draw a frame.
4. Think about ways of naming important features, for example:
   (a) write names on the drawing, or
   (b) number each one and write the list at the side.

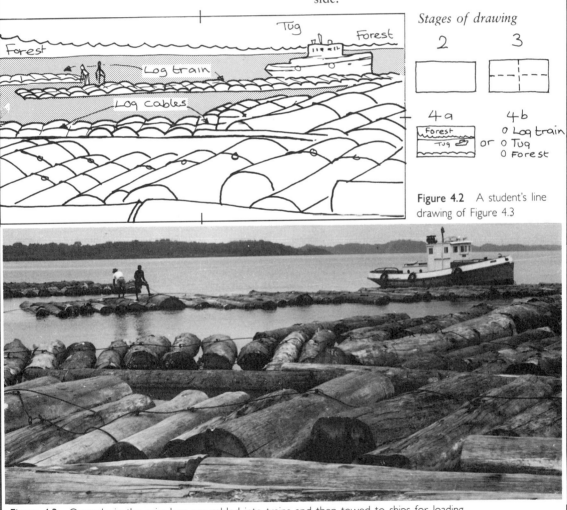

Figure 4.2 A student's line drawing of Figure 4.3

Figure 4.3 Owendo in the rain: logs are cabled into trains and then towed to ships for loading

34 The New Africa

petrol-filling stations, tailors, and shopkeepers.

But will the forest still exist in the year 2000? In theory governments control the amount of timber to be cut, and follow a strict re-planting programme. Read the extract from the newspaper report on the Ivory Coast to find out what can happen (Figure 4.4).

Forest cover blown

In less than a generation the Ivory Coast's tropical rainforest is likely to be only just a memory. In 25 years 12 million hectares of woodland has shrunk to less than 4 million.

A recent study by a team of experts from Abidjan's Institute of Tropical Geography concludes that "In the face of different consumers of land, the total disappearance of the wooded area is expected by the end of the century". This will have profound economic consequences and hidden climatic effects:

- The drop in wood sales will leave a gap in the country's trade balance.
- The jobs of some 30,000 employees in the timber industry are in jeopardy.
- As the forest recedes rainwater either evaporates or runs off at an accelerated rate.
- The savanna zone in the north advances.

The forestry companies open up vast tracts of previously inaccessible forest. Land-hungry farmers then clear out all vegetation to plant cash crops like coffee and cocoa.

The Ivory Coast's forests have been sacrificed over the past twenty years on the altar of immediate profit. The future generations may well have to pay the price.

Figure 4.4 The disappearing forests

The main reason for giving so much attention to the forests of tropical Africa is that they are a *renewable resource*. When the mineral oil, manganese or iron ore deposits are used up, Gabon and other countries will still have an income earner in their forests – provided it has not been exhausted.

How healthy is the economy of Gabon? To find out more about it look at the summary table.

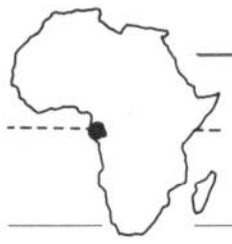

Gabon	Capital: Libreville, 252,000 (1975)
Exports $2,000 million Imports $726 million	Visible trade balance $1,274 million surplus (1983)

Export commodities (1982)	%	Export partners	%
Petroleum	81	France	30
Timber	8	USA	26
Manganese	5	Netherlands	7
Agricultural products	3	Italy	5

Study the list of exports and answer these questions.

1 How many products or groups of products are listed? Would you describe it as well diversified (that is, having many *different* exports)?
2 Does it include manufactures as well as products of mines, forests and agriculture?
3 Note the first three export products. What percentage of cash income does each provide?
4 How does the total value of exports compare with the total value or imports?
5 If a country exports more than it imports it has a trade *surplus*. If it imports more than its exports it has a trade *gap*. Which does Gabon have? This is called the *balance of trade*.

Using trade summary tables to compare African countries

Look at summaries for the Congo and the Central African Republic and compare them with Gabon.
- Compare the Gabon export total with the other two countries. What is the main source of Gabon's wealth?
- Which country is the odd one out because its principal export is not petroleum?
- Which of the countries does not have a visible trade surplus?
- What export commodities do the three countries have in common? In what ways are Congo or Central African Republic similar to Gabon?
- Which of the countries has the most diversified range of exports?

There is another source of information which can be used to compare countries and find out basic facts. Look back at Figure 3.5 (page 29) and find the GNP Figures. The GNP for Gabon was $2,950 million in 1983, quite a lot more than that for the Congo and the Central African

Gabon: a forested country in equatorial Africa 35

	Central African Republic	Capital: Bangui, 350,000 (1982)
Exports $115 million		Visible trade balance
Imports $138 million		$23 million deficit (1983)

Export commodities	%	Export partners	%
Coffee	33	France	64
Diamonds	24	Belgium/Lux	16
Timber	20	USA	3
Cotton	6	Israel	3

- A landlocked state heavily dependent on agriculture.
- Diamonds are exported but their importance is declining.

Republic, which are larger both in area and population. How does the GNP per head compare?

Trade figures should be used with caution. They do not show all the activities going on in a country but they do show how a country earns its foreign exchange.

For more information on country summary statistics and an explanation of these terms read the fact box on page 36.

	Congo (Brazzaville)	Capital: Brazzaville, 422,000 (1980)
Exports $1,066 million		Visible trade balance
Imports $650 million		$416 million surplus (1983)

Export commodities	%	Export partners	%
Petroleum	90	USA	51
Timber	5	Italy	21
Diamonds	2	Spain	10
		France	10

- Like Gabon most of the land is densely forested, and timber is an important export.
- It has a frontage to the Atlantic Ocean of only 100 miles.
- Offshore oil has been the main source of wealth since 1979 and has transformed a visible trade deficit to a surplus
- It is among the top ten African countries on the basis of income per head.
- It provides important rail links for Gabon and Zaire river transport to Pointe Noire and is at a crossroads for transit trade.

Figure 4.5 Logging in Nigeria:
- The background shows unaltered rain forest.
- Note the size of the felled logs measured against the men. They all wear protective 'hard hats'

Facts: How to get country information from summary tables

Look back at Figure 3.5. For each country it shows:
- How large the country is, the number of people living there and the date when it became independent.
- The population density, that is the number of sq km to be 'shared' by the people who live there.
- The percentage of population living in the larger towns and cities.
- The gross national product (GNP) and GNP per person

Part 2 of the book includes a trade summary for many of the 48 countries in Africa, and sometimes a short written description. Each table shows:
- The principal export commodities and their destinations
- The value of exports and imports and the visible trade balance
- The name and population of the national capital

Make sure you understand the following.

Exports

The trade summary tables show the *main export commodities* earning foreign exchange and the *main trading partners*. These can vary from year to year depending on the world market price and production difficulties. Details of import commodities are not given as these are fairly similar for nearly all African countries. It is just the overall quantities that vary. The trade figures are in *US dollars* for the most recent available year. There are some inconsistencies due to changing exchange rates and gaps in the statistics.

The visible trade balance

The exports and imports of commodities and manufactures form the *visible trade* of a country. If exports are more than imports there is a visible trade *surplus*. If imports are more than exports there is a visible trade *deficit*.

There are also some '*invisible*' movements of money which do not enter the visible trade figures. The tourist industry is a good example. If a country offers fine scenery, safaris for wild game, or just sunshine, people come to visit and spend money. Another example is the money sent home to their families by people working in other countries. Thus the *visible trade balance* does *not* give a full picture of a country's trading position.

Population and area

The *density* of population is calculated by dividing the number of people by the area of the country (sq km). For example, a population of 2.8 million divided by an area of 56,000 sq km gives a density of 50 persons per sq km (Togo). The average density gives a rough idea of how empty or crowded a country is, and of the pressure on resources. These average figures hide great variations between one part of a country and another. The number of people living in a country is both a resource and a responsibility.

Wealth

Economists measure the 'wealth' of a country by adding up the total value of all goods and services that are produced and sold in a country, both within it and abroad. The total amount is called the *gross national product* (GNP), and when divided by the total population it gives the average product per person (that is, *per capita*). This can be used to compare the relative wealth of one country with another even when total populations differ. Clearly, in the real world the wealth is not shared out equally among the population. The GNP also leaves out items such as food grown and eaten without being bought. The GNP figures are in *US dollars for 1983*. The per capita calculations are based on estimated populations for 1983. A more recent total population figure is shown where available.

Chapter 5 Zaire: an equatorial giant

> **Key words**
>
> Size, diversity, transport problems, multi-national companies, economic islands.

Zaire is the third largest country in Africa. Only Sudan and Algeria are larger and Zaire is twice the size of the Republic of South Africa. It is 10 times the size of Britain. Yet only about 30 million people live there, an average density of 13 people to every sq km. So this vast equatorial country is only thinly peopled.

The River Zaire is the second largest river in Africa after the Nile. Most large rivers in Africa are shared by several countries. The Nile (over 6,000 km long) flows through Uganda, Sudan and Egypt. The River Niger (4,000 km) flows from Guinea through Mali, Niger, and Nigeria. Figure 5.1 shows that the boundaries of Zaire almost match the river basin but it also includes part of the Great Rift Valley and related lakes and volcanic mountains on the east.

The life of Zaire focuses on the river which forms the most important system of inland waterways in Africa, a total of 11,000 km with its navigable tributaries. Railway lines have been built to by-pass unnavigable sections of falls and rapids (see Figure 5.3). There are about 250 different ethnic groups in this huge and varied area, forming 70 tribes or nations. Along the river and its tributaries, Lingala or 'river speech', has developed and is becoming a national common language. French is the main administrative language.

It would be wrong to think of all Zaire as a densely forested equatorial country, though parts are. There are three main landscapes:
- the forest
- the wooded savanna (both of which are cut by great river valleys)
- and the highlands in the east

The forest is found on either side of the equator in the northern half of the country. To the south and north there are wooded plateaus, deeply cut by rivers with forested valleys.

The centre of the basin was formerly a great lake where sediments were deposited. The bedrock of the continent (the Basement Complex)

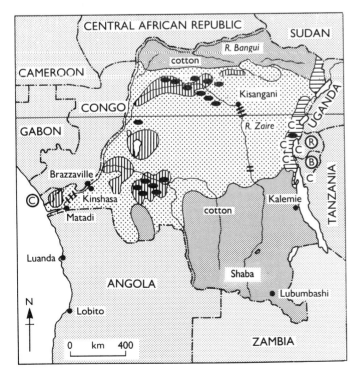
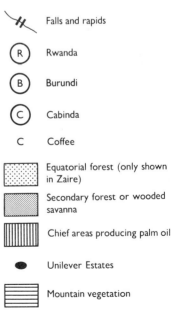

Figure 5.1 Zaire: landscapes and land use regions

which they cover therefore comes to the surface only as a broad rim, but an important one because it is in these ancient rocks that the most valuable minerals are found. On the east, the highest part is formed of volcanic rocks bordering the Great Rift Valley (see Chapter 6, page 44).

Zaire lacks good access to the sea. Not only has it a very small coastline (about 40 km long, though it shares part of the ocean estuary with Angola), but the river has to cut through the western part of the highland rim in order to reach the sea. Over 300km of rapids separate the capital, Kinshasa, from the sea and its shallow draught port at Matadi and a deep water port at Boma. While inland transport alternates between river and railway, the last 300km to the sea has to be done by rail or road.

Rural development and agriculture

Zaire's agricultural potential is huge and well diversified. Recent agricultural policy has shown a shift away from export crop production towards
- food production for local consumption, especially maize, cassava, rice, sugar and fish (from fish ponds as well as rivers)
- agricultural products which can be used as inputs for domestic industry, for example, maize, for milling, brewing and cattle feed, cotton for textiles, and palm oil for domestic use (cooking, palm wine) as well as for export.

Most people live in the countryside and earn a living from the land. But 80 per cent of the income of the country comes from mineral production.

The farms produce many different crops. Only some of them are *market* crops that can be sold for cash income; the rest are domestic crops required by the local people. We often hear of coffee, cocoa and rubber as market crops for export. A less well-known crop is the fruit of the oil palm. It is especially valuable because it is both edible and an industrial product, used in soap and margarine manufacture.

Oil palm products

Details about the oil palm are given in the fact box (and also Chapter 7). Its 'home' area is the wet, humid forest of west and central Africa. The wild tree has been domesticated and improved. Zaire has plantations of oil palm, but it also produces oil from wild palms growing in natural palmeries. In such a situation it is important that:
- African farmers growing market crops on their own holdings should not be at a disadvantage
- both African farmers and large plantations

Facts: Palm Products

Climate and Cultivation
- Temperature: constant high temperatures 18°C minimum
- Altitude: not above 1,000 m
- Rainfall: over 1,500 mm, well distributed
- Sunlight: abundant, so cannot compete with tropical forest
- Soils: wet and heavy. Grows well on banks of rivers, swamps and lakes where soil is too wet for rain forest vegetation, and in forest clearings.
- Small farms: much of produce comes from tall (10 m) semi-wild trees which mature in 8–10 years. Fruits are harvested November–April by hand.
- Estates: particularly in Zaire, Ivory Coast and Benin, and state farms in Ghana. Seedlings are planted out in rows and trees are smaller and mature after 3 years. Inputs (labour for weeding, sprays and fertilisers), and yields greater.

Processing
Oil is extracted from the fleshy fruit by pounding and boiling or by a hand press or hurried to a local oil mill to prevent deterioration. Estate products are processed using sophisticated machinery (see Figure 5.2). Kernels are cracked and pressed for oil.
Three different products:
- Palm oil – used widely locally as unrefined oil for cooking but also processed in factories into margarine, soap and candles.
- Palm kernel oil – refined and used in margarine, toilet soaps, cosmetics, and glycerine.
- Palm wine – sap is 'tapped' like rubber, fermented, and drunk.

African production
Africa produces 23% of the world's palm products. Nigeria is Africa's largest producer (50%) followed by Zaire (10%) and Ivory Coast (8%).

Zaire: an equatorial giant

Figure 5.2 Trucks of oil palm fruits are lined up to be pushed into the steam ovens prior to the extraction of oil

should grow good quality products.

Standards must be maintained both in growing and in pressing. There must be plenty of small oil mills distributed in such a way that farmers in the countryside can easily take their fruit to be processed.

The advantages and disadvantages of plantation agriculture

Zaire offers a good example of palm oil production on both smallholdings and plantations. Think about these questions:

1. Is it easier to apply scientific methods on plantations, for example, the selection of seed, the application of fertiliser and the control of disease and weeds?
2. Where are yields likely to be higher and trees bear fruit earlier and produce over a longer period?
3. Do plantations interfere with local customs of land holding and farming?
4. Do small farms avoid high overheads and expenses and grow food crops on which the farmers' families can live until the cash crop matures?
5. Does the single crop of a plantation exhaust the soil?
6. Can pests and diseases more easily run through a single-crop plantation than through different crops on scattered holdings?
7. Why is one crop farming economically risky?

Unilever: an example of a multinational company

In Zaire, palm oil and kernels are produced from small farms and Unilever plantations.

- The stocking of plantations from wild palm groves began early this century in the then Belgian Congo. By the mid-1980s there were 10 estates.
- This multinational company also has rubber, cocoa and tea estates in Zaire making the total up to at least 15.
- These estates form part of the Unilever global 'family'. Worldwide, Unilever has about 800 businesses in 5 continents.
- There is an enormously wide range of activities in addition to plantation products: chemicals, plastics, paper and packaging, meat products, detergents (soap liquids), foods, drink of all kinds, supermarkets, building and construction firms, transport and shipping.
- Multinationals often open up areas with little or no infrastructure. They introduce efficient, cost-effective production.

> 'Do multinationals help or hinder development?'

- But Unilever is reducing the estates in Africa because it is cheaper to produce palm oil in South-east Asia.
- If an area becomes less productive or there are labour problems multinationals move out to another part of the world. They are not interested in supporting the local economy.

People argue that multinationals hinder rather than help developing countries. They are in business to make profits. What matters is the balance between:
- a country's gain in jobs, roads, bridges, air strips, schools, clinics, etc., as well as its share of the profits
- and the money that is sent out of the country back to the shareholders. These are the people who have lent (invested) the capital. They, in turn, would take their money out of Unilever if it did not make profits for them.

Timber production

Nearly half of Zaire is covered by forest which has so far been little exploited. As in Gabon, if it is used carefully, the forest could form an important renewable resource when the minerals run out. Production and exports of logs have

40 Central Africa

increased greatly since 1980 but are still only one-quarter of Gabon's.

The mineral wealth of Zaire

Mineral resources supply over 80 per cent of the Republic's income. In 1982, Zaire produced 27 per cent of the world's industrial diamonds, 26 per cent of the world's cobalt, and 6 per cent of the world's copper. It is the 9th largest producer of tin in the world, and Africa's 4th largest gold producer. In 1984, Zaire's most important source of foreign exchange was copper (38 per cent of exports), and second was mineral oil (18 per cent) from the country's small offshore oilfields. Water power potential is possibly the greatest in the world – 13 per cent of the world's total and 40 per cent of the estimated total for Africa.

Figure 5.3 shows that Zaire may be divided into two clearly separated parts.

• In the west and centre the rivers are thickly marked to show that they are navigable. Their gradients are gentle, the beds are free from rapids and there is adequate depth for navigation at

Zaire	Capital: Kinshasa, 2.4 million (1976)
Exports $1,726 million Imports $1,121 million	Visible trade balance $605 million surplus (1982)

Export commodities	%	Export partners	%
Copper	38	Belgium/Lux	53
Petroleum	18	USA & Canada	16
Diamonds	13	France	8
Cobalt	12	Italy	7
Coffee	12		
Zinc	3		
Tin	2		

most seasons of the year. This is the lower 'sedimentary' section of the basin.

• In the north-east, east and south-east (the 'Rim'), the land is higher, the river beds are more rocky and irregular and less good for shipping; but the steep slopes make power sites possible.

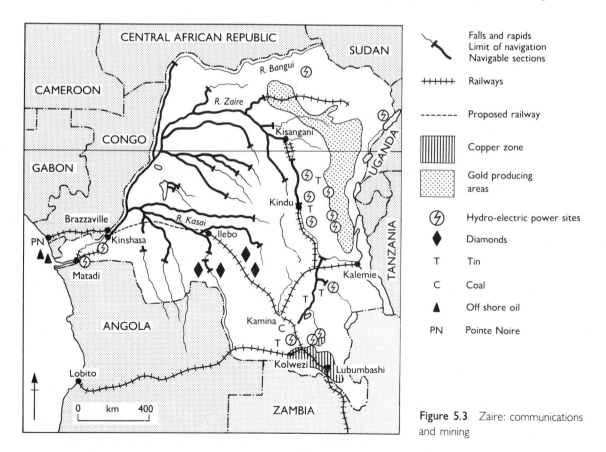

Figure 5.3 Zaire: communications and mining

Figure 5.4 A copper mine near Kolwezi, Zaire.
Note: • ore quarry in foreground
• processing plant in middle distance
• housing estates in far distance

This is the part where the bedrock of the African continent comes to the surface. Look for a great arc of gold-bearing rocks in the north-east and the copper zone in the south-east. The two main zones are sometimes called the Koperzone and the Tinzone. Nearer to the centre of the country, river gravels in South Kasai provide an important source of industrial diamonds.

The most fully developed areas are those of the Eastern Provinces, especially Shaba in the south-east. Other well developed and settled areas are those near Kinshasa (Lower Zaire), and the high lands of the eastern borders which have some of the most beautiful lake and mountain scenery in the world, and a height which makes living and working more pleasant in an area so near the equator.

The Shaba Koperzone and cobalt production

The Koperzone of Shaba forms a population island. It joins with the population island of the Copperbelt of Zambia to form one of the most important mineral centres in Africa between the Sahara and Johannesburg (see Figure 5.4). Together they produce about 13 per cent of world copper. (For a study of copper mining and production in Zambia turn to page 127).

There are two main problems:
• The need for good communications emerges from the size of the country.

• The problem of a lack of skilled manpower and organisation is experienced by all rapidly developing countries. It is more serious in Zaire because of political unrest and uncertainty.

Zaire can only develop its potential if there is stable government, smooth organisation, skilled Zairean workers, good management and reliable communications.

For nearly 50 years Zaire has been the world's biggest producer of cobalt, a by-product of copper. Cobalt is a rare and strategically important mineral. So far there is no alternative to its use in alloys in jet engines and the aerospace industry. It stands very high temperatures, wears more slowly, and resists corrosion. The problem is that cobalt is associated with copper lodes in Shaba, so to get it you have to mine more copper – of which there is a world *surplus*.

The three major problems facing copper and cobalt production in the Koperzone are:
1 The fall in copper prices and its replacement by other cheaper materials.
2 The need for skilled manpower, good organisation and political stability.
3 Lack of cheap and reliable transport to the sea from Shaba (see Figure 14.3, page 123).
• The shortest, Benguela, rail route through Angola to Lobito has been effectively closed since 1975, but international moves have been made to reopen it:
• The Beira route through Mozambique is kept open by Zimbabwe troops protecting the line.
• About half of Zaire's copper is exported through the Republic of South Africa – a long, expensive, and politically unpopular route.
What are the problems of an all Zaire route?

Size, communications, and uneven development

What does the future hold for Zaire? The main problem is still one of size, communication and uneven development. Each of its provinces is as large as a European country. It is 1,600 km from Lubumbashi to Kinshasa, the political and financial capital. Before air transport it took about 9 days by river and rail. Kisangani is 1,300 air kilometres from Kinshasa or 10 days up and 5 days down by water. At such a distance the regional (provincial) capitals must have a large measure of independence.

Use Figure 5.1 and Figure 5.3 to find the economic islands.
1. Bas (Lower) Zaire which has towns, ports and villages powered by electricity from the Inga Dam. It also has a huge cement works, sugar refinery, water schemes, and countless small industries.
2. Shaba Koperzone in the south-east. It is powered by electricity from Inga.
3. The agriculturally rich highland rim bordering Uganda, Tanzania, Rwanda and Burundi in the east. This area is potentially one of the world's great tourist areas with lakes, active volcanoes, mountain gorillas and prestige lodges and hotels in game parks.
4. The mineral islands: the Kasais (diamonds), and the tin areas.
5. Other agricultural export islands, mainly in the west and related to river transport.

Note that these sites are peripheral, that is they are distant from the centre of the country.

The zones of agricultural, mineral and general economic development are separated by vast distances where there is not enough money to keep up roads, telegraph lines or river transport, which is often laid up for repairs.

Rapids make it impossible to use the River Zaire for its whole length. Where rapids or falls interrupt the river, railways have been built to avoid them. But this means that goods are repeatedly shifted from rail to ship then back to rail. The train past the rapids may only go once a month. Large river steamers can be thought of as floating hotels — for the first class passengers. For the other 1,500 travellers packed in with all their luggage and food they are more like a floating village. A stop may be several hours or 20 minutes. Local people wade out to sell food. Steamers have huge headlights which act as searchlights at night to navigate the river in the total darkness of the forest.

Zaire today has a good manufacturing base compared with many African countries. It produces a wide range of consumer goods for the domestic market: food, drinks, tobacco, textiles, footwear, wood products, metal manufactures, chemicals and cement. But industries are mainly carried on in the cities such as Kinshasa and Lubumbashi, and in Lower Zaire (near the river mouth) and other regional centres. Some rural areas are still isolated and almost unchanged. In between the economic islands and small towns Zaire remains a country of hundreds of traditional villages.

Zaire is capable of becoming one of the most prosperous countries in Africa. So what stops it? Some of the problems have already been noted:
- transport difficulties
- the need for skilled manpower
- the drop in world prices affecting exports
- crippling loan repayments on large capital projects such as the Inga dam hydroelectric project

Some of these problems are man-made, some natural.

Eastern Africa

Chapter 6 The plateaus, coastlands and mountains of East Africa

> **Key words**
>
> Rain-makers, cattle management, interface, farm systems, consolidation, import substitution, site, situation and function of towns.

Kenya, Tanzania, Uganda, Rwanda and Burundi are the countries of the high plateau of east Africa which occupies nearly half the width of the continent near the equator. The plateaus stretch mile after mile for hundreds of miles, interrupted by occasional hills, monotonous, baked brown or red in the dry season, green after the rains.

The height of the land has been important in 3 ways:
- High altitude reduces temperature and this ensures that the East African countries do *not*

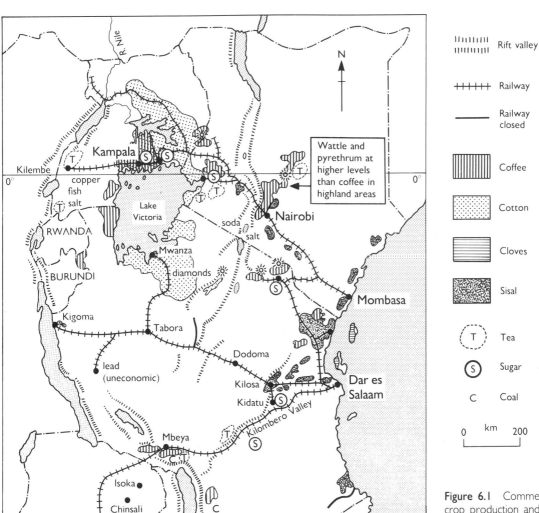

Figure 6.1 Commercial crop production and communications in eastern Africa

have the type of climate usually described as equatorial. The climates are so varied that a wide range of crops can be grown, both tropical and temperate (see Figure 6.1).
- The higher land is often formed from volcanic materials poured out when rifting occurred and the great volcanoes were formed. These materials weather into fertile soils, far better than most plateau soils, and are sought after by farmers.
- Good climate and spectacular scenery attracted Europeans to settle and farm permanently.

In no other part of Africa is there such a variety of landscapes within such a relatively short distance:
- in the west are some of the most spectacular mountains and active volcanoes, and sheer fault cliffs bordering the lakes of the western Rift Valley (Zaire, Rwanda, Burundi, Tanzania)
- there are the two arms of the great Rift Valley
- there are limitless plateaus, broken by great isolated cones of extinct or dormant volcanoes – Kilimanjaro, Kenya, Elgon and a host of others
- there is a varied and beautiful coast bordered by coral reefs
- and of course there are the man-made landscapes of cities, farms, game-parks.

No wonder 'whites' wanted it, and for half a century thought of it as theirs. Before them the Omani Arabs and others settled the coast from the Asian Gulf States helping to create the Swahili culture.

Figure 6.2 Salt being loaded at a crater lake. Mineral salts dry out in crater lakes. Slabs are prised from the bed and floated to the shore on 'rafts', stacked to dry, and then distributed all over Uganda

This chapter has 3 sections:
1. East African climate, weather and land use.
2. People and land in East Africa.
3. The growth of towns.

Figure 6.3 Kigezi, south-west Uganda. The steep slopes are terraced to reduce soil erosion. This area is typical of high rainfall mountain areas. The volcanic soils are very fertile and carry high population densities

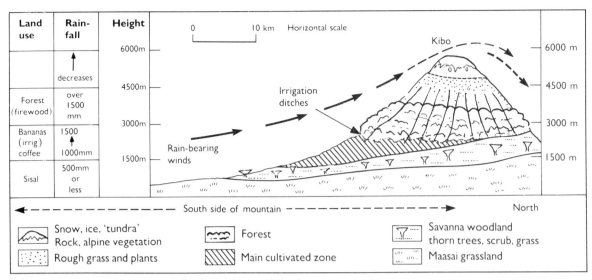

Figure 6.4 Changing land use with height on Mount Kilimanjaro

East African climate, weather and land use

The equator passes through the east African region, yet much of it is neither equatorial nor truly tropical. No part has the typical average annual rainfall of 2,000 mm spread throughout the year with average temperatures near 25°C, and a small daily range.

Mountains as rain makers

The most important single factor is the height of the land. East African land rises from sea level on the east coast to 4,000 m or more. As the height changes so does the climate. Just as important as the *seasonal* rainfall change is the way rainfall *totals* increase with height. Mountains rising above the plateaus are rain makers throughout tropical Africa. This is true even during the dry season. For example, the Maasai grasslands south of Mount Kilimanjaro are parched and yellow in August: yet 50 km away on the middle slopes of the mountain there are clouds and thunderstorms and a rich green land of bananas and coffee farms.

Look at Figure 6.4 to find out what happens on Mount Kilimanjaro. The land heights are shown on both the northern and southern sides of the mountain.

1. How high is the Kibo summit of Mount Kilimanjaro?
2. What is the height of the lowest plateau land shown on the south?
3. Find the forest symbol. Does it go right across the mountain?
4. Roughly how many metres is the forest belt from low to higher levels?
5. Find the main cultivated zone. It does not continue right across the mountain: which side is it on and why?
6. The land-use belts are related to different annual rainfall totals. Name the land use with (a) the lowest rainfall and (b) the highest.
7. Does the rainfall change evenly right to the top of the mountain?
8. What occupies the very top of the crater of Kibo?

Rainfall unreliability and farming

The feature of the weather and climate that makes most difference to the people of east Africa is rainfall *un*reliability.

There are places on or near the equator that are almost rainless. This is clearly shown on the Rainfall Uncertainty map (Figure 6.5).

The reasons for this are complex. The south-east and north-east trade winds tend to blow parallel to the coast and neither gets very far inland. Some winds come from dry areas such as the wastes of Saudi Arabia.

46 Eastern Africa

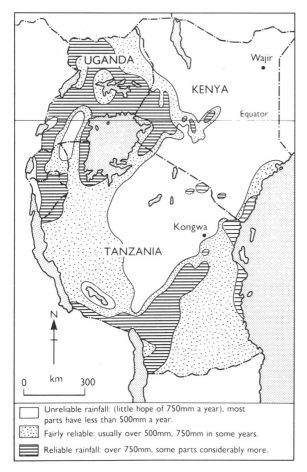

Figure 6.5 Eastern Africa: rainfall uncertainty

About half of the land in east Africa has rainfall totals so low and unreliable that farming is at risk. A second problem is that the dry zone drives a wedge hundreds of kilometres wide between the coast and the interior.
- In Tanzania the dry zone occupies a huge area in the middle of the country separating productive areas.
- In Kenya an even larger part has unreliable rainfall; but the area of good agricultural land, though small, includes some of the best land in Africa.

People and land in East Africa

The east African countries, including Rwanda and Burundi, have a total population of over 60 million, twice the population of Zaire. There are 4 main studies in this section:

- Dry zone pastoralists.
- Small coffee farms in Uganda.
- The importance of the highest land: the Kenya Highlands.
- Commercial farming: combining large and small-scale sugar production.

Dry zone pastoralists

In many parts of Africa the dry lands are pastoral areas grazed traditionally by stock owning families, for example, in the Sahel, in Botswana and other savanna areas and desert margins. All pastoralists have to know where they can find water and grazing during long dry seasons. So they are migrants, or 'nomads' but use a regular rotation of sites. Perhaps best known in eastern Africa are the Maasai (see Figure 3.2).

Maasai Rangelands
The Maasai people make a success of living in one of the most difficult environments in Africa. Not only is the average rainfall between 250 and 500 mm – a very low annual total in a hot tropical country – but there are long dry seasons during July, August and September, and during December, January and February. Three months of drought is a great strain on cattle who need 5 litres of water every day, or at least every other day. A way of life and special skills have been developed to deal with this.
- The people of the driest areas move regularly each season towards higher wetter pastures, and only use the Rift Valley pastures during the rainy periods. In Kenya this is from March to June and October to November.
- The cattle keepers know which bulls adapt best to arid conditions and use them to serve cows.
- They mate animals so that the births do not come at a time when there would be little water or grazing.
- They know about tsetse flies and the cattle sickness, nagana. Tsetse flies also have a seasonal migration; the Maasai move in when the fly moves out. If they have to take animals across a river into a fly belt they do so at midday. This is a 'safe' time because tsetses die after only one hour of strong sun.

The traditional way of life (see Figure 6.6) is no longer as strong as it was and even the diet is changing. Many people enjoy *posho* (a millet porridge) and other foods. Some of the pastoralists are already adopting a different way of life as teachers, farmers, veterinary officers, nurses and mechanics, and some in public life. Nevertheless

The plateaus, coastlands and mountains of East Africa 47

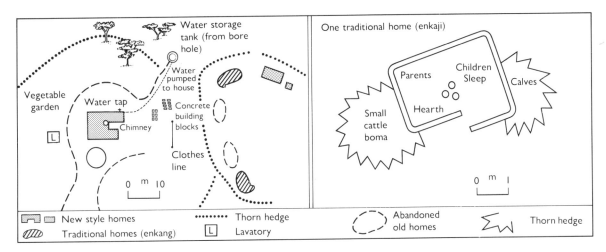

Figure 6.6 Changes in Maasai homes

there is probably less change among the Maasai than among many African peoples.

Future use of the dry lands
Governments everywhere are providing water points, veterinary help, slaughter houses, cold stores, and marketing. The greatest problems are still overstocking — which causes overgrazing, destroys vegetation and leads to soil erosion — and water. In prolonged drought there is not even enough water for people. Small-scale methods such as windpumps may come to the rescue.

> 'Two thirds of the land of Africa is at risk from unreliable rainfall. But good land management can transform the dry lands'

Windpumps at Wajir (see Figure 6.5) bring water to the surface from a depth of 45 m.

Some problems
• The underground water table can be lowered too far. Careful control is needed.
• Windpumps supply good drinking water: they cannot usually provide enough water for *regular* irrigation.
• Grazing is ruined around water points if it is trampled by too many animals.

Part 3 describes large irrigation projects that could transform the dry lands. But all these changes alter the traditional way of life. In some places that can be irrigated there is competition for the same land between the pastoralists and the newcomers. This can be a zone of conflict. Such a place where different users of the same land confront each other is called an *interface*. This is a good word to remember because it can be used in other situations, for example, the interface between an urban and rural area, where developers want to take over good agricultural land.

Small independent farms in Uganda

On mountains, rainfall and temperature change with height. On smaller hills, soil, moisture and wind conditions vary down a slope, making each part suitable for different crops and land use.

A flat topped hill and farmed slope in Uganda: a matching exercise

1 Study the drawing, Figure 6.7A in practical work box 3, showing a number of smallholdings near Kampala, Uganda, and the transect, Figure 6.7B, below it. (Find out what a transect is from practical work box 6, page 93.)
2 Look for the 6 different uses of the land by matching the numbers on the drawing and on the transect.
3 Estimate the proportion that each might occupy if you walked from the top of the hill down the slope to the valley swamp. This is a rough estimate of the proportion of land used in each way.
4 Answer the following questions.
• What rock capping makes it difficult to grow

48 Eastern Africa

Practical work box 3: making a landscape sketch out of doors

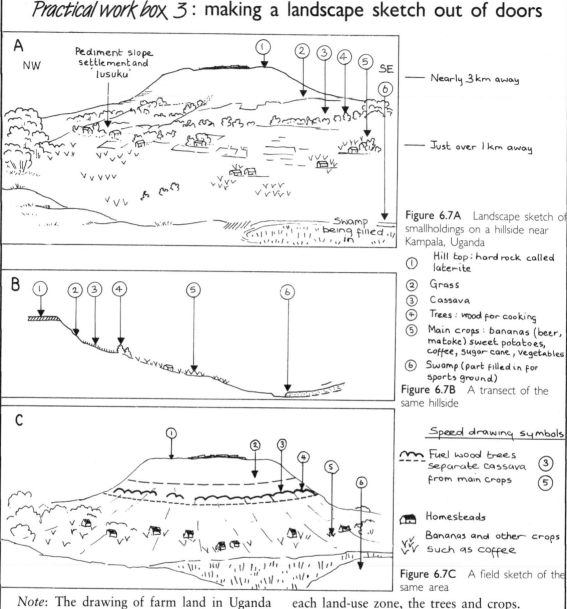

Figure 6.7A Landscape sketch of smallholdings on a hillside near Kampala, Uganda

① Hill top: hard rock called laterite
② Grass
③ Cassava
④ Trees: wood for cooking
⑤ Main crops: bananas (beer, matoke) sweet potatoes, coffee, sugar cane, vegetables
⑥ Swamp (part filled in for sports ground)

Figure 6.7B A transect of the same hillside

Speed drawing symbols

- Fuel wood trees separate cassava ③ from main crops ⑤
- Homesteads
- Bananas and other crops such as coffee

Figure 6.7C A field sketch of the same area

Note: The drawing of farm land in Uganda (Figure 6.7A) and the transect (Figure 6.7B) are used for the matching exercise on page 47. The field sketch (Figure 6.7C) shows how to simplify a drawing made out of doors.

• First look back to practical work box 2, page 33, to see how you made a simple line drawing from a photograph.

• You decided the most important things to mark on Figure 4.2: the log trains, the open sea, the forested shore. Here, in Figure 6.7 they are: the shape of the hill, the position of each land-use zone, the trees and crops.

1 Decide the speed drawing symbols to be used (suggestions are shown alongside C).
2 Quarter the landscape to help get the proportions right as you did for Figure 4.2.
3 Try to estimate the distance for yourself (suggestions are marked at the side of A).
4 Add labels. In this example only one label is written on sketch A (pediment slope settlement) and land use is referred to by numbers 1 to 6. You can add labels round the frame if you leave space.

crops at position 1?
• Crops are not grown at position 6. What could it be used for?
• Give reasons why the homesteads and main cropped area are found at position 5.
• Why are the trees at 4 important?
5. The transect can be drawn on the blackboard in class and discussed.

Each holding cuts across different types of land so it is of more use to the farmer than a plot that is suited to growing only one kind of crop.
• He can choose where to grow crops needing more or less water or sun or protection from wind, and can leave the poorer or more exposed ground for hardier crops.
• He can grow perennials, annuals, and cash crops such as coffee as well as food crops.
• He has a varied diet and different crops need attention and marketing at different times, so his labour is spread out. A year that is not very good for one crop may prove better for another.

Such a farmer lives well and makes money. A similar pattern of land use change on hill slope is found in many parts of Africa (see Figure 6.8).

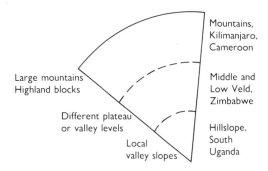

Figure 6.8 Scale wedge: changes down a slope at different scales

---

Facts: Coffee

There are 2 main varieties:

ARABICA	ROBUSTA
Climate and cultivation	
• Temperatures: 19–23°C	• Higher: 20–26°C
• Altitude: 1,500–2,300 m	• Lower: 1,000–1,500 m
• Rainfall: 800 mm minimum	• Higher minimum: 1,000 mm, constant humidity
• Soils deep; well-drained	• Lighter but moisture holding
• Seedlings and cuttings nursery grown in shade	
• Planted out at 12 months and pruned to 2 m, weeded and fertilised	
• Harvesting starts after 3 years. Berries (cherry) ripen 8–10 months after flowering and are hand picked	
Farm systems	
• Grown on large estates and smallholdings	• Grown mainly on smallholdings
Processing	
• Cherry wet processed and fermented to remove pulp	• Cherry mainly dry processed: sun-dried on farms
• Beans sun-dried then 'parchment' skin removed in mills	• Sent to hulleries for removal of pulp and parchment
• Beans roasted, graded, packed	• Beans roasted, graded, packed

African production and export
Africa produces and exports about 22 per cent of the world's coffee
Ethiopia, Uganda, Cameroon, Kenya and Ivory Coast are the most important producers

Farm systems

When books say 'coffee is grown on estates in Kenya and on smallholdings in Uganda' such a statement covers two quite different *systems of farm production* and the details of the lives of many people. These 'shorthand' terms leave out details: the skill or the problems of the grower; why he prefers to grow coffee to cotton or vice versa; how he keeps the pests out of his crop; whether his family pick the crop or whether he hires labour; who markets it and who pays for transport; whether the price of coffee stays level or drops so much that he thinks of switching to cotton; whether his land is too wet to grow good cotton, and so on. Yet these are the things that matter in people's lives. If the grower earns enough he can hire labour and his children can go to school. If his crops bring in less, he must use his children as pickers and they lose their schooling and someone else loses a job. Countries and people need to spread their assets through many different crops and enterprises rather than having 'all their eggs in one basket'.

The policy of varying one's crops (and other economic activities) is called *diversification*. Not only is it a form of insurance to grow a variety of crops, it is often a means of improving the diet.

Small mixed farms grow a wide range of domestic crops for home use or local markets. They also try to grow at least one market crop, in this case coffee.

Use the coffee fact box to find out more about growing coffee and the tea fact box to find out about growing tea. The photograph, Figure 6.9, was taken on a tea estate in Kenya.

Facts: Tea

Climate and cultivation
- Temperatures: not too high, therefore
- 1,000–2,200 m altitude
- Rainfall: 1,300 mm minimum, prefers 1,600–2,000 mm
- Soils: acid, damp, deep, well drained
- Leaf cuttings nursery grown for 18 months in polythene bags (sleeves) to 25–30 cm, then planted out in contoured rows
- Pruned at intervals to form a branched 'plucking table' first at 40 cm then at 60 cm level
- Plucking can start at 2½–4 years then at intervals of about 2 weeks throughout the year up to 50 years. Only 2 leaves and a bud plucked
- Pruned every 3–4 years down to new level (April)

Farm systems
- Grown on large estates of 400 ha or more and by outgrowers (smallholders) concentrated near factories.

Processing
- Leaf hurried to factory, dried in controlled conditions
- Crushed to free juices, fermented, dried and cooled
- Graded and packed into plywood chests for export
- 5 kilos of green leaf produce 1 kilogram of made tea

African production and exports
Kenya is the 6th world tea producer and 4th exporter. Malawi is the 11th producer and 6th exporter. Tanzania, Mozambique and Zimbabwe are also important.

Figure 6.9 Tagabi tea estate, Kericho, Kenya. This photograph summarises estate tea production: rain, the smooth plucking table, factory buildings, homes for workers

The importance of the highest land in Kenya and other parts of tropical Africa: geography and politics

In Europe low land is usually more important for agriculture than high land. In some parts of tropical Africa it is the reverse.

In Kenya nearly one-fifth of the land is over 1,500 m above sea level. There are two 'wings' of high land on either side of the lower land of the Rift Valley stretching west to Lake Victoria. This is the economic heart of Kenya (see Figure 2.3).
• The higher land receives higher and more reliable rainfall in what would otherwise be a semi-desert area.
• The height of the land makes it a comfortable living place for Europeans as well as Africans and Asians.
• Higher land in East Africa is often the result of volcanic outpouring and volcanic rocks develop into fertile soils.
• The high land is beautiful as well as productive.
• The variations in height provide opportunities for growing a greater variety of crops. Some of the advantages and the problems of Kenya are based on these facts.

At one time the Kenya Highlands were known as the White Highlands and were reserved for European farming. Three of the most densely settled areas of Kenya adjoin the former European farmlands, and land hunger and sometimes actual hunger provoked demands for land. This was one of the reasons for the outbreak of the movement called 'Mau Mau' during the 1950s.

It is always difficult to state a case fairly for both sides, especially in a few lines. The following points present some of the views about the rivalry for land in Kenya:
• When farmland is 'fallowed' in the tropics it looks empty.
• Immediately before the main European takeover of East Africa and the building of the railway from Mombasa to Kisumu on Lake Victoria in 1902, there had been years of drought and crop failure, cattle disease, and tribal wars.
• Many people died and others moved from the plateau to the better watered slopes of Mount Kenya and other highland areas.
• In many African societies land is available to those who need it and becomes 'yours' if you cultivate it. If cultivation stops it reverts to the clan. It is the product of the land that is really 'owned' by the farmer when he cultivates. The Kikuyu had traditional clan land, though some of it had previously been in the hands of other tribes, including the Maasai who once hunted on the slopes of Mount Kenya.
• When the African population increased gradually the people asked for their farmland back; but this was resisted by the Europeans.

Since independence in 1963 land in Kenya is no longer specially 'reserved' for the use of Europeans. But there are still large and well-run company-owned estates as well as thousands of smallholdings, both here and in the other East African countries.

Land resettlement and rural transformation in Kenya

For many years, and especially since 1955, there has been a policy of land improvement in the East African countries. Scattered plots of land, for example, of the Kikuyu, were *consolidated*, that is, surveyed and regrouped to make one farm holding. As a result, yields of maize increased until the dry years of the 1970s.

Development plans aim at
• Improving the quality of life in the countryside

Figure 6.10   The stages of land development in Kenya

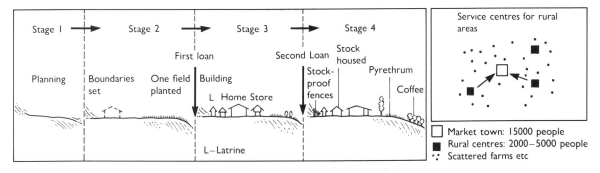

by helping ordinary people to share in all aspects of development. 'The key to success in any plan is the participation of the people'. It includes *Harambee* or self-help.
• The development of *service centres* to serve rural areas.
• Organised marketing so that farmers can plan ahead and be freed from exploitation by small middlemen traders.

The resettlement programme in the Kikuyu area was spectacular, but it is only one of the areas of East Africa where land improvement is going on. The stages of land development are shown in Figure 6.10.

Ninety per cent of the people of Africa still live in the countryside. So the transformation of rural life has to take priority. In most African countries the ownership of, or the right to share, land is a basic social insurance. Increased productivity from the land will provide the schools, clinics and homes that country people want.

It is essential that high agricultural standards should result in high yields in the one-fifth of Kenya that can produce good crops, because nearly two-thirds of the land cannot. The produce of the land is vital to Kenya not only for the people living there but also because half of Kenya's exports are agricultural products.

> *Facts*: Sisal
>
> *Climate and cultivation*
> • Temperatures: high but very tolerant
> • Altitude: 0–1,800 m (very adaptable)
> • Rainfall: 750 mm desirable but tolerates wide range
> • Soils: tolerates wide range
> • 'Bulbils': (seeds from long flower pole) planted in rows
> • Leaves: hand cut after 3 years and up to 10 years
>
> *Farm systems*
> • Mainly grown on large estates – quick transport to factory essential. Grown near railways and on the coast.
>
> *Processing*
> • Decortication: flesh stripped off leaves
> • Fibres washed, dried and bleached in sun. Brushed, graded, baled for export
> Sisal fibre is used for twine, sackcloth, matting.
>
> *African production*
> Kenya is the second largest world producer (13%) and Tanzania third (10%).

Commercial farming

The first large-scale farming in East Africa produced mainly primary products for export: sisal, tea, coffee, cotton, sugar. Soon huge ranches, wheat and dairy farms were added on land high and cool enough for temperate crops. Since independence there have been changes:
• Some of the former white farms have been divided up and the land used for resettlement.
• Processing and manufacturing of farm products has resulted in secondary products, ready for sale locally or for export.
• Some products that were imported are now manufactured in East Africa. This not only creates jobs but saves scarce foreign currency. This is called *import substitution*.

Large resettlement schemes like those at Mwea and Ahero gave land to landless farmers. But there has not been enough good land to go round. Small farmers find it difficult to compete with the large commercial farms. Large-scale development often disrupts local communities and may take men away from their families. Thus all African countries have been looking for a farm system that successfully combines the large and the small scale. The Mumias sugar study shows that small-scale and large-scale enterprises can be effectively combined.

The sugar estates and outgrowers' system at Mumias, Kenya

In the early 1970s Kenya was spending about $US20 million a year on imported sugar. This was about 4 per cent of the visible import total. Kenya wants to develop industries, make more jobs, and save foreign exchange to reduce its trade deficit. It has done this successfully for some industrial products, for example the Pan African paper mill at Webuye. The problems are:
• how to find the capital and invest the huge amount of money carefully
• how to choose a site that will not take land from existing small farms, and
• how to gain the 'know-how' needed for success

The government asked the company Booker International to prepare a pilot study. They

The plateaus, coastlands and mountains of East Africa 53

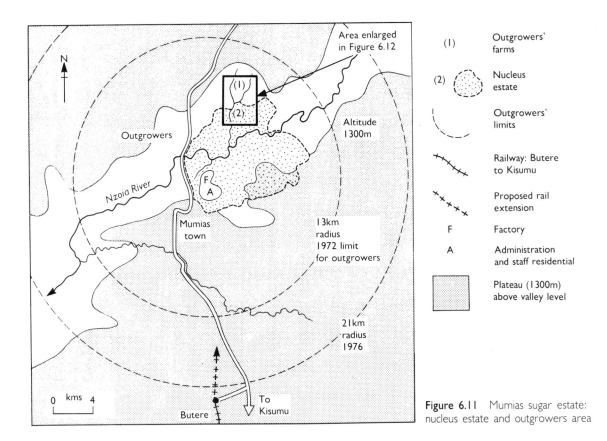

Figure 6.11 Mumias sugar estate: nucleus estate and outgrowers area

consulted the local people and came up with a proposal that would suit both sides. Here are some of the points made by the company:
- We need a huge area – 4 to 5,000 hectares – and local workers.
- The best site is the Nzoia River near Mumias.
- Few people live in the valley because it floods. We could control the floods.
- Local farmers living above flood level could become "outgrowers" and have their cane processed at the estate sugar factory.

Here is the point of view of the local people:
- We don't want to lose our farms.
- The Mumias area is very densely settled.
- We could grow some sugar as a cash crop and still grow our own food crops.

Now find out more about the system that was agreed by studying Figure 6.11 and Figure 6.12. Figure 6.11 shows the situation of the Mumias estate.
Find:
1 the nucleus estate: factory, administrative area
2 the valley of the Nzoia River and tributaries
3 the outgrowers 13 km radius (1st stage) circle and 21 km circle
4 the area covered by Figure 6.12

Figure 6.12 shows a small part of the nucleus estate and a group of outgrowers' plots.
Find:
1 the large sugar estate blocks in the southern half
2 the separate but linked outgrowers' sugar plots in the northern half
3 'chain' 23: each 'chain' has a reference number. Chain 23 has 11 plots, each plot belongs to a different family. Only the sugar plot is shown, but each holding has land for food crops going down to the stream.
4 the track from the road used by heavy tractors going to service sugar chain 23.

Do you understand why outgrowers' sugar plots cannot be scattered, but must be linked?

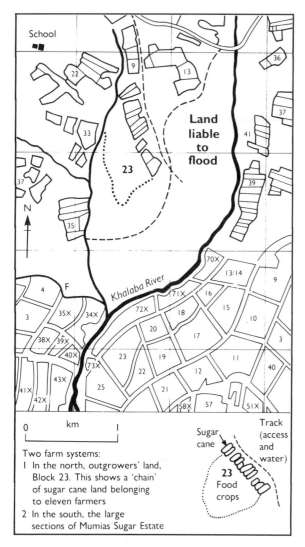

Figure 6.12 Details of sugar plots at Mumias. The area shown is marked on Figure 6.11

The remarkable success of Mumias is shown by the following:
- Mumias now produces 62 per cent of Kenya's sugar.
- The nucleus estate sugar covers 3,400 ha. Outgrowers' sugar covers 32,000 ha (10 times as much).
- There are more registered outgrowers within the 13 and 21 km radius circles by in-filling, and there are now some farms beyond the 21 km circle. The result is that the total number of outgrowers is about 30,000.
- By the early 1980s, Kenya's sugar imports had dropped to less than $US 2 million, about 0.1 per cent of the import bill.

Facts: Sugar Cane

Climate and cultivation
- Temperature: average of 20°C, no frost, therefore
- Altitude: 1,500 m upper limit
- Rainfall: 1,500 mm spread over 9 or more months each year, or irrigation
- Soil: well drained
- Stem cuttings (setts), dipped in fungicide/insecticide are hand planted in succession to give all year round supply
- 3 crops hand cut in a cycle of about 65 months:
- Crop 1: cut after 20 months growth
- Crops 2 and 3: regrowth from later shoots ('ratoons') cut at intervals of 18–20 months
- Ground cleared and given 8 months' rest before replanting

Processing
- Hurried to factories (48 hours maximum), weighed, chopped and crushed
- Juice treated with lime and sulphur, heated and clarified, boiled and crystals separated from molasses
- Dried and bagged

African production
South Africa produces 27%, Egypt 12%. Mauritius, Sudan, Kenya, Zimbabwe, Swaziland are also important.

1 Look at the practical work box on page 55 to find out how to 'simulate', that is, act out, a real life situation.
2 Use the discussion points above as a 'trigger' for the role play exercise.

You could have a discussion like this about any large development.

The growth of towns

When people lived in the countryside and depended on their own produce they had little to sell and little money to spend. Now they sell surplus goods at informal markets. Some of the money is spent on school fees. The rest buys clothes, shoes, bicycles, radio sets and other articles, some of them from overseas. Thus even

Practical work box 4:
practical decision making: how to 'simulate', that is, act out, a real life situation and reach an agreement

In the real world policy decisions often have to be hammered out. It happens in small or large businesses, in government offices, in parliament, in schools, in the family. Acting out a *role*, that is playing your part in a 'real life' situation, teaches you a lot. Perhaps your future success will depend on examination results. But it will also depend on your ability to get along with other people and your ability to make good decisions.

How to set up a public inquiry based on sugar growing at Mumias
1 Everyone should quickly read through the Mumias study, pages 52–54.
2 Make a list on the blackboard of all the *people* involved. These are some of the roles to be played:
- two government officials
- two representatives of the company
- local chiefs
- health care officers
- local farmers

There may be others.
3 Choose people to play the individual roles, and groups to represent farmers, villagers, etc.
4 You also need to choose two people to act as reporters (chairmen). They will keep order at the inquiry, record the main points of the discussion and make sure you reach a decision.
5 Try to think of other things that should be considered (for example, transport links, a new railway line?).
6 Uses the *trigger* information: the Mumias exercise, the two diagrams and the discussion points.
7 Each group should *list* its point of view, ready for a representative to argue its case. (These record cards can be kept as part of the study.) *Keep it simple*.
8 Set yourselves a *time limit* for:
- group discussion and preparation (5 minutes?),
- the inquiry itself

Role play
Role play can help you to understand what to expect in different jobs. Mumias is an agro-industrial situation. Role play in other situations would give you some idea of what it is like to work in an industry such as mining, or as a planner in an overcrowded city.

To be effective you have to:
- learn to sort evidence and organise ideas and facts
- learn to present your point of view clearly and politely
- learn to negotiate. A good decision is one that satisfies all parties so is more likely to last.

Why give time to a 'game'? Does it teach you anything?
- It helps you to find out how the real world works.
- It helps you to understand how people react and how they change their opinions.
- It helps you to adjust to different situations: to face up to your own or your community's problems, and be sympathetic to other people's needs.

---

in the countryside small towns are springing up which can collect and transport crops and supply the goods and services people want. This is happening all over Africa.

The smallest service centre of this kind grows up at the point where people can easily meet, an open space where roads meet.

In East Africa many of the shops are called by the Swahili word '*duka*'. Because of this some of the new townships and shopping and service centres are called duka towns. Mukono in Uganda is a good example.

56 Eastern Africa

Mukono, a duka township

Study Figure 6.14.
Make a list of all the features and services shown on Figure 6.14, grouping them into categories concerned with
- relief and drainage
- transport
- economic activities
- welfare provision
- other amenities

Some important items are not shown, for example, there is usually a post office, a church or chapel, a school, a local government maintenance centre for building and repairing roads and bridges, and sometimes an agricultural service centre. People use the Post Office Savings Banks for most of their transactions and they are now found in towns and villages over much of Africa. If you live near Mukono bring the map up to date.

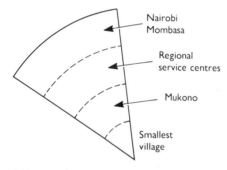

Figure 6.13 A scale wedge to show the relationship between settlements of different sizes

The study of towns: Mombasa

The coastal towns and ports of Africa are the starting points for important rail and road routes inland. These are good places to store, process and distribute imported goods and to provide services that are needed. If towns are well placed, they can expand their activities, and attract more people from other areas.

As a town grows, parts of it become used for different types of activity such as the shopping districts, the residential areas, the industrial areas, business district, or port. These are functional areas concerned with special work or services, and are easily recognisable. There are also areas of mixed development where people live and work and where there are small workshops and markets.

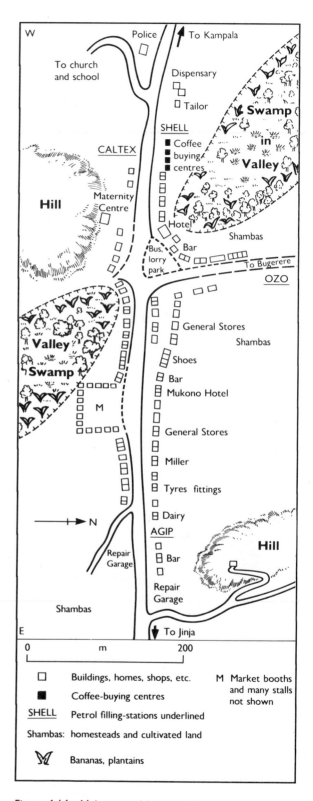

Figure 6.14 Mukono: a duka township

The plateaus, coastlands and mountains of East Africa 57

Practical work box 5: making your own maps: Mukono as an example

You can make your own simple map of a road, part of a village or farm as you walk along. Use the following method:

1. Practice before you go out.
• Copy the map of Mukono, Figure 6.14, by drawing it step by step on a piece of paper or on the blackboard.
• Draw the map as if you were walking through Mukono from south to north. Follow the instructions on Figure 6.15. They tell you what to do at each stop-point, A, B, C, D.
• It should look like Figure 6.16.

2. Out of doors.
• The first time you make a road traverse out of doors choose a simple one with a safe place to walk and stand.
• Don't take your exercise book out – it might rain!
• Make a simple clipboard. You need a piece of stiff cardboard or thin plywood and a strong clip to hold a piece of paper in place. Cut a groove around the top of a small piece of pencil and use string to tie it to the clip.
• You can map side roads or farms in the same way as on Figure 6.16.
• Count shops (as you did market stalls at stop point C). This can be broken down into a simple census of different kinds of shops, filling stations, etc.

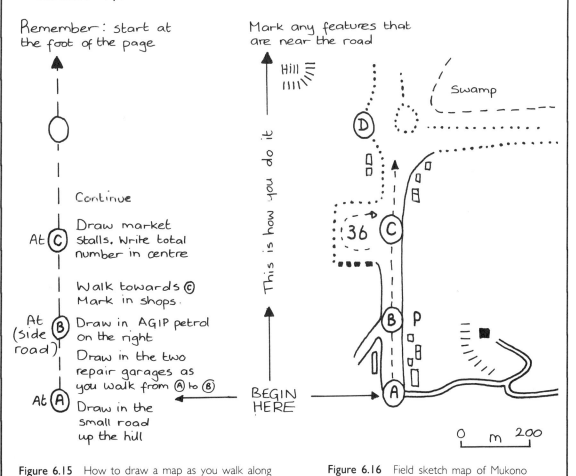

Figure 6.15 How to draw a map as you walk along

Figure 6.16 Field sketch map of Mukono

58 Eastern Africa

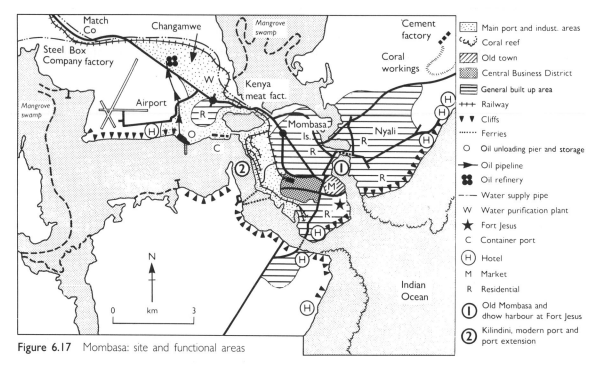

Figure 6.17 Mombasa: site and functional areas

Look at the map of Mombasa (Figure 6.17) and its key to find:
1 the old dhow harbour, the anchorage for the early sailing ships from the Persian Gulf (1)
2 the old town, guarded by Fort Jesus, close to the dhow harbour
3 the modern port at Kilindini, where there is much more space for shipping, port facilities, and rail access (2)
4 other functional areas and special facilities in and around Mombasa: residential, industrial, etc

Similar types of functional areas are found in nearly all large towns and cities. Their importance varies with the main functions of the town, for example a port, a national capital, or a mining centre. The main types of functional area can be found on the maps of Mombasa, Lagos (page 96), Nairobi (page 185), and elsewhere in the book. Use this list to check them.

Mombasa provides useful material for noting the general *method* of studying towns. They are often studied under 3 headings: *site*, *situation*, and *function*. So far we have discussed only function.

• *Site* The site of a town is the actual land on which it has grown up. Most towns have an original site and as they grow the site extends. In the case of Mombasa the original site was on the eastern side of the island where there was a protected anchorage. The narrow entry through the coral reef and bluffs gave good shelter and protection both from weather and brigands. Fort Jesus was built by the Portuguese to give added safety though they did not keep it very long. The narrowness of the harbour entry was one of the factors that made it inevitable that a larger harbour would one day be needed. Hence the two port areas 1 and 2. Today the site of

Name	Function
• central business area (or district, often called the CBD)	offices, major government, civic and cultural buildings; shops, hotels
• residential areas	living areas, including housing, schools, shops, etc, to serve the residents
• industrial areas	heavy manufacturing and processing plants (e.g. oil refinery) light service industries (e.g. printing) and port and railway areas.

The plateaus, coastlands and mountains of East Africa

Mombasa covers the whole island and overflows onto the mainland.

- *Situation* The situation of a place concerns its general position in relation to other places. Mombasa is on the east coast of Africa 80 km north of the boundary with Tanzania. It has links with other places by sea, air, and land, especially the interior of Kenya, and Uganda, Rwanda, and Burundi further to the west. 'Situation' also concerns the efficiency of communications and the importance of the places with which there is contact, and the area over which the influence of a centre extends. People starting industries or services want to know the possible market where they can sell their goods or services. The approximate limits of the area a port or town serves for different purposes can be shown on a map. The area of influence of a town is sometimes called its 'field', borrowed from the idea of the 'magnetic field'.
- *Function* The functions of a town are the activities that bring in money and give its population their jobs. The jobs in a port like Mombasa, relate to the handling and transport of goods and the processing of products. Look at Nkana–Kitwe (page 127) to see the difference in a mining centre where a mineral product, copper, forms the economic base of the town. All towns also serve the day to day needs of the population for housing, shopping, education, administration, etc. This is called the *tertiary sector*.

Kenya	Capital: Nairobi, 1.1 m (1984)
Exports $1,034 million Imports $1,336 million	Visible trade balance $302 million deficit (1984)

Export commodities	%	Export partners	%
Petroleum products	27	UK	13
Coffee	26	West Germany	11
Tea	14	Uganda	10
Fruit & vegetables	7	Netherlands	5
Cement	4		
Sisal	2		

- There is development of small holder tea and sugar farms.
- The Tana river schemes will provide a major source of power.
- The new paper mill at Webuye saves importing paper and foreign exchange.
- Geothermal power from 0l Karia in the Rift Valley feeds electricity into the grid (see page 194).

'More war, more want'

Uganda	Capital: Kampala, 458,000 (1980)
Exports $229 million Imports $278 million	Visible trade balance $49 million deficit (1981)

Export commodities	%	Export partners	%
Coffee	97	UK	26
Cotton	1	USA	22
Tea	1	Netherlands	20
Copper	1	Japan	14

- Uganda has been ruled by fear and force for nearly two decades.
- The economy dislocated, and public services and infrastructure have deteriorated.
- Uganda is now one of the poorest countries in Africa.
- There is potential for tourism, for successful tea and coffee growing, and for copper/cobalt production.

Figure 6.18 An African businessman-farmer filling his water cistern on his vegetable farm near Mombasa

60 Eastern Africa

Tanzania (including Zanzibar) — *Capital:* Dar es Salaam, 880,000 (1980)

Exports $688 million
Imports $1,038 million

Visible trade balance $350 million deficit (1981)

Export commodities	%	Export partners	%
Coffee	39	UK	21
Diamonds	23	West Germany	15
Cotton	12	USA	11
Cashew nuts	8	Netherlands	5
Tea	6		
Sisal	3		
Cloves	2		
Tobacco	2		

Former President Nyerere's policy of *ujamaa* or 'familyhood' was aimed at reducing the gap between the living standards of town and country. The plan was to set up villages where schools, clinics, water supply, and medicare could be more easily provided than for scattered rural people, and to farm the land cooperatively, so that the scarce inputs (tractors, agricultural advice) could be shared. Sadly these aims were hit by:
— a fall in crop yields due to drought
— drops in the market prices of export crops
— increases in imported oil prices
— the cost of Tanzanian troops in Uganda
— Tanzania's dependence on foreign loans

Rwanda — *Capital:* Kigali, 118,000 (1978)

Exports $143 million
Imports $198 million

Visible trade balance $55 million deficit (1984)

Export commodities	%	Export partners	%
Coffee	65	Belgium/Lux	17
Tea	18	Uganda	12
Tin	9	Italy	2
Pyrethrum	1	France	1

Burundi — *Capital:* Bujumbura, 172,000 (1979)

Exports $81 million
Imports $184 million

Visible trade balance $103 million deficit (1983)

Export commodities	%	Export partners	%
Coffee	84	West Germany	34
Tea	7	Italy	4
Cotton	4	Belgium/Lux	4

- Rwanda and Burundi together form a mountain area reaching 2,400 m
- Climate– vegetation zones range from savanna to mist forest
- They have the highest average population densities in Africa.
- The shortage of land results in erosion and falling yields.
- They are the least accessible countries in Africa
- However there is:
— fertile volcanic soil, good rainfall, and the potential for tourism
— scope for hydroelectric and geothermal power, and supplies of natural gas and peat
— deposits of vanadium, nickel, copper, and platinum

Western Africa

Chapter 7 The Guinea lands: coast and forest landscapes

> **Key words**
>
> Farmed forest, parastatals, micro-climate, intercropping, large and small-scale farm systems, primary products

In the area of eastern Africa just studied the main contrast is between lowland and highland in the same latitude. In the Guinea lands, situated between the Gulf of Guinea on the south and the Sahara desert to the north, the main contrast is between
- the coastal forests and
- the northern savannas further away from the equator.

This is a simple, memorable pattern but it hides a good many variations. There are 4 main natural landscapes as one moves northwards from the coast:
- coastlands and forests,
- wooded or tree savanna,
- dry savanna with thorn trees and more grass,
- desert.

Each zone runs parallel to the coast. Each results from a distinctive climate. The forests have high temperatures, high rainfall (2,000 mm or more) and high humidity for most of the year. The northern savannas have a dry season increasing to 5, 6 or 7 months with distance from the coast and total rainfalls of only 500–1,000 mm a year. A journey inland from most of the important coastal towns such as Lagos or Abidjan to Timbuktu would traverse all these zones (Figure 7.1).

Between November and April when the overhead sun is in the southern hemisphere and the climatic zones move south, Saharan conditions prevail over much of West Africa. At this time winds blow strongly and steadily from the northeast carrying fine Saharan dust. This is the harmattan season. First the sun hazes over but the weather is still hot and sticky because the dry, dusty air is well above ground level. Suddenly the harmattan arrives and the skin is dry and crisp.

Only the south-west Guinea coast and southern Nigeria escape the harmattan. But even at Warri (see Figure 9.6) in the extreme south of Nigeria, December, January and February have a total of about 100 mm of rain instead of 500 mm in the three months, which average equatorial weather would give.

As the weather belts move north again in the northern summer, equatorial conditions take

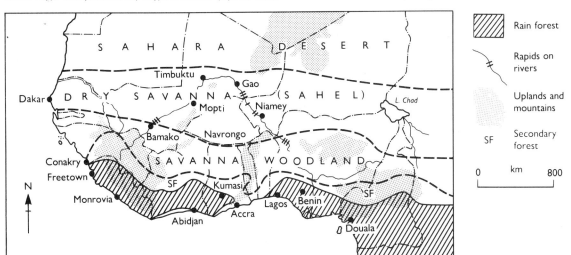

Figure 7.1 Vegetation and landscape zones in western Africa

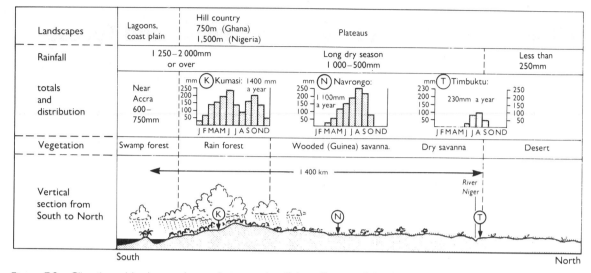

Figure 7.2 Climatic and landscape changes between the Guinea Coast and the Sahara Desert: a transect

over. Most places within 150 km of the coast have 1,500–2,000 mm or more of rain each year. The savanna lands whose dry season rainfall is only 50 or 75 mm in 5 or 6 months receive nearly 1,000 mm or more between May and September when Guinea Coast conditions prevail (Figure 7.2).

Most of the countries of the Guinea coast stretch inland across the forest zone and the tree savanna, into the dry savanna. Only in the extreme south-west on the monsoon coast are there countries such as Sierra Leone and Liberia with little savanna. There are, however, many variations within the 2 major types, especially when the land rises to a thousand or more metres as in Cameroon.

The West African studies in the next 3 chapters are:

Chapter 7
- Farmed forest
  - rubber in Liberia
  - timber and plywood, Sapele in Nigeria
  - oil palm bush, Nigeria
  - cocoa forest in Ghana and Ivory Coast.

Chapter 8
- Traditional savanna farming in northern Ghana: problems of the long dry season.
- Seasonal life on the River Niger, and contrasts in water use:
  - a 1940s irrigation project
  - Mopti, a river port
  - Kainji dam
- The future of the Sahel.

Chapter 9
- Agricultural towns and service centres in West Africa.
- Lagos: a multi-functional city.
- Abuja: a new capital city.
- Mineral oil in the Niger delta.
- Nigeria: boom and slump.
- West Africa: the future, ECOWAS.

Forest landscapes in West Africa

Areas of true rain forest in West Africa are limited.
- They need over 1,800 mm of rain well distributed through the year
- They only survive where there is little interference from farmers
- Most areas have 1,300–1,800 mm of rain with 2 peaks separated by spells of drier weather, and some of the trees are deciduous (Figure 7.2)
- Although the rainfall varies and with it the height and type of tree, the 3-storey appearance remains (Gabon, page 32).

Farmed forest

In the past, people who lived in the forest supplied almost all their needs by gathering fruits, edible roots, pliant stems for ropes, cutting trees and canes for building, and using herbs, pods and leaves for flavouring and 'relish' sauces. They scarcely disturbed the forest.

Today there are many changes in the forest zone even though the main weapons for clearing may still be fire and a cutlass. Under family

The Guinea lands: coast and forest landscapes 63

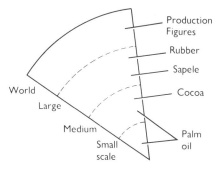

Figure 7.3 A scale wedge: four West African forest production studies

clearing methods the stumps of large trees are not removed even by burning. The land is put down to food farms under the shade of the remaining tall trees. Thus the forest becomes food-farm forest, or cocoa forest, or oil-palm bush.

For much of this century improved technology for cutting, hauling and shipping timber has made inroads on the forest. Often only the most valuable timber trees were taken out. But farmers moved in and the 'high forest' ecosystem was broken down. (Refer back to Gabon, page 32)

Four studies show how the forest can be farmed, that is, can be productive and earn money without being destroyed. Two are large-scale extensive methods, 2 are small-scale intensive systems. They are:
- Rubber extraction in Liberia.
- Sawn timber and plywood at Sapele, Nigeria.
- Oil palm bush in Nigeria and
- Cocoa forest in Ghana and Ivory Coast.

All 4 are based in the countryside. Three studies are of primary production, that is, the first stage of production before manufacturing. The other, Sapele, is both primary and secondary, because it produces wood but manufactures it into a wide range of secondary products such as furniture and doors.

The first two, rubber and timber, need heavy capital investment for their large-scale operations. The second two show how local people can get the best out of their own small farms (see Figure 7.3).

Note that some of the largest cities in West Africa are in the forest zone: Ibadan (Nigeria), Kumasi (Ghana), and the coast cities, Lagos, Abidjan, Monrovia and Freetown.

Rubber production in Liberia

Rubber is a good example of a wild product that has been 'domesticated'. Much time was wasted collecting rubber from scattered trees in the equatorial forest. Planting on estates:
- saves time
- ensures a high quality product

But trees are not large enough to tap for the first 6–7 years so a very large amount of capital must be invested before any money is earned by the company. That is why big business is often described as *capitalist*. Countries invest money in a similar way in state enterprises (parastatals) which will take some years to become profitable.

The Firestone Company in Liberia
- There are 70,000 acres of plantations in 5 estates (13 million trees).
- There are also 140,000 acres of trees farmed by 5,000 independent small producers.
- Rubber is exported in its raw state (see rubber fact box) so no further manufacture of tyres or other products takes place in Liberia.

Figure 7.4 Rubber production in Liberia. Overhead cables are used to move cans of latex to the weighing stations

Figure 7.5 How rubber trees are tapped

Facts: Natural Rubber

Climate and cultivation
The hevea rubber tree grew wild in Brazilian forests, but is now planted on large estates. It needs:
- high temperatures – 21°C average at least
- high rainfall – 1,750 mm minimum per year, evenly spread ('no rain, no sap')
- low altitude – below 1,000 m
- deep well-drained soils

'Farming'
- Seedlings are planted out in rows 50 to the acre and are ready for tapping after 6–7 years.
- The outer bark of the tree is cut so that the sap runs (see Figure 7.5).
- This white sap, called latex, is collected three hours later.
- Each tree is tapped on alternate days.
- Trees are rested after about 6 months tapping.

Organisation
Estates: Most successfully grown on large estates where collecting can be concentrated. Each tapper is in charge of a fixed number of trees (see Figure 7.4)
Small farms: Small farmers sell their produce to big estates/companies

Processing
At the estate factories latex is:
1. Treated with acid to coagulate (thicken) it so it can be rolled into flat crepe rubber sheets and smoked, or
2. Concentrated from 30 to 60 per cent rubber by removing water so it can be exported in liquid bulk.
Both these products are then exported.

African production
Liberia, Nigeria and Ivory Coast produce and export most of Africa's rubber.

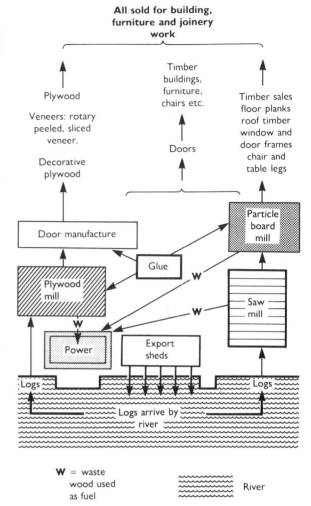

Figure 7.6 Sapele timber and plywood, Nigeria: a flow diagram of processes and products

Timber and plywood at Sapele, Nigeria

Sapele is the biggest industrial timber complex in Nigeria. It is possibly the largest wood based industrial operation in Africa. While Liberian rubber is a primary industry – it processes the rubber for export, but does not manufacture it into secondary products – Sapele combines both primary and secondary production. So the company AT&P, as it is called, farms timber from government concessions and ships it by launch, raft or tug to the Sapele mill.

1. Look at Figure 7.7, a line drawing of an aerial photograph of the Sapele mill.
2. Use the key to name four important products.
3. Find the two largest sheds numbered 1 and 2 and check their position on the flow diagram, Figure 7.6. These are the two major operations.
4. Check the position of the power house. Electricity and steam are generated from the mill's own wood waste and diesel oil.

The Guinea lands: coast and forest landscapes 65

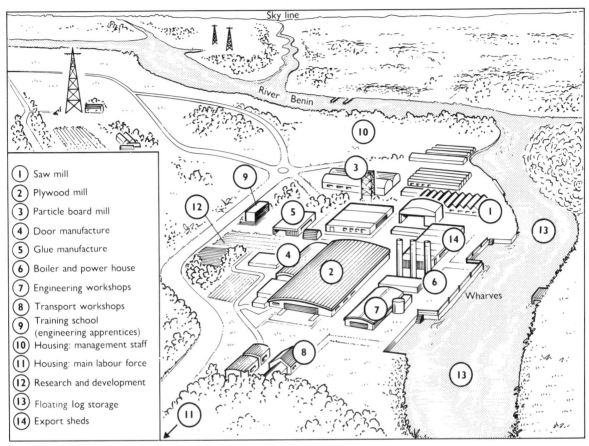

Figure 7.7 African timber and plywood, Sapele: a line drawing based on the area shown on Figure 7.8.
1. The point of this exercise is to learn to recognise features shown on air photographs.
2. It helps us to know something about the layout of a big industrial plant and its parts.
3. It is *not* to be learned 'by heart'.

5 The third important unit is the particleboard mill. Particleboard is made from saw and plywood waste and small timber. It is chipped into flakes, mixed with glue and pressed hard into boards.
6 Study the flow diagram, Figure 7.6, to see how these three operations give rise to a range of market products. Make sure that you understand all the stages.
7 Have a good look at the line drawing, Figure 7.7, again. Check the key and note the position of all the units that have to work smoothly together if production is to continue.
8 Use the line drawing, Figure 7.7, to identify as many buildings as possible on the photograph, Figure 7.8, especially the river wharves, log storage, and the site itself in the middle of the forest.

The timber industry: conservation and exports

The rainforests of the world are shrinking at such a rate that some places, including West Africa, may have little left by the year 2000 (see Gabon, page 34).

Now governments and large companies like AT&P are trying to preserve and renew the forest by the following methods:
• Trees in forest concessions are not cut until they are ready. Only mature trees are taken out.
• In some places a 30-year logging cycle is followed, that is, trees are left to grow for 30 years after a spell of selective cutting.
• Valuable plywood trees, like mahogany, take between 50 and 200 years to mature. Researchers are looking for alternative trees to these slow growers. One, called gmelina, from south Asia is

Figure 7.8 An air view of the Sapele mill

one of the world's fastest growers. It matures in 7–10 years compared with 50 years for West Africa's own trees. It is suitable for door and window grames, or grinding for particleboard.
- A reforestation experiment for gmelina has been planted near Benin City – to be ready 1992 onwards.

Governments are taking measures to stop illegal felling and damage. Although Nigeria produces 20 per cent of Africa's timber it does not export much, partly because timber exports have been banned in the past, partly because Nigeria has a huge home market. The government is now encouraging controlled cutting and some exports.

Soil fertility in the forest zone

The lush growth of the rainforests of Africa suggests a fertility which often does not exist. Removing whole stands of timber or clearing by burning for farming, sets off a succession of changes:
- The newly exposed soil is baked by the sun.
- This helps to break down the soil structure; such land is easily eroded by heavy rain.
- The layer of leaf-litter normally protects the soil from pounding by rain and baking by sun.
- When the clearing returns to bush the secondary forest is thinner so that the leaf-litter layer, which makes the humus in the new soil, is less.
- The leaves on the new bush growth are fewer or smaller and therefore they discharge less moisture into the atmosphere. It becomes a small area of different climate – *a micro-climate* – in which fewer seedlings of some of the original trees survive, so that these species are not renewed. The result of all this is that the secondary forest is different from the primary forest.
- Soil-exhausting weeds quickly take possession of a clearing returning to bush.

The Guinea lands: coast and forest landscapes 67

Figure 7.9 Multi-cropping in the forest zone. A good African farm in the forest always looks overcrowded. Discover from the text why this is good. Yam vines climb the sticks in the background, but the crop with the large leaf in the foreground is cocoyam. The plant in front of the men is cassava

The fertile soil of the forest is usually only a thin skin of about a foot in depth. The forest giants have only shallow roots, the reason why they so easily blow down in strong winds.

Family farming methods involve less disturbance of the forest than other systems, especially if permanent farming is achieved by the introduction of semi-permanent tree crops of the tropics, such as cocoa, coffee and rubber. This saves the labour of constant clearing and burning, and shades and protects the soil, because these trees are introduced as *part* of the forest system, not instead of it.

At first sight an African food farm looks a muddle (see Figure 7.9). But it is an advantage for the ground to be completely covered by different crops, grown together or *inter-cropped*. They ripen at different times so the ground is never really bare. If the season is a drier or wetter one, it will at least suit some of the crops, whereas a one-crop planting might mean a complete loss.

Farming the oil palm bush

Palm oil production shows farming by selection. If the oil palms, which are fairly resistant to fire, are protected when land is cleared and burned, they increase in number in the following years. Gradually land becomes a natural concentration of oil palms.

The oil palm provides important export products. It provides a staple item of diet as a cooking oil and in stews and other dishes. It is also tapped or cut for palm wine. It is therefore grown throughout the forest zone wherever the heavy rains and rather swampy soils that it prefers are found. (See Figure 9.8, page 90.) These areas of secondary forest, or oil-palm bush, are often some of the most densely populated parts of the west coast of Africa (see Figure 7.10). In southern Nigeria village surveys suggest that there are as many as 100 to 200 people to a square kilometre. There are many family compounds and clusters of villages, each with its market, lorry park, shops and often an oil mill for pressing the oil-palm fruits.

The land farmed by a family is only a hectare or less. It is difficult to estimate exactly how much land is used because more distant parts are not farmed every year; and if the family owns 80 or 90 oil palms they will be picked in rota.
- Immediately round the compound the land is manured with refuse and dung, and yams, maize, cocoyams and other crops are planted under the palms.
- The value of the manured ground is shown by the weight of yams which average 1 kilogram, while those from 'outside' farms weigh only half a kilogram.
- The 'outside' farm land is not manured. It is planted with yams or cassava for 1 or 2 years and then has two, three or four years of fallow. Almost every year a new piece of bush is opened up while the old land rests.
- Over half of the food used in the family is bought.
- Unfortunately, crop yields are falling because the pressure of population on the available land has shortened the resting period; fertility is no longer restored to the land before it has to be used again.

Although there are oil-palm plantations, for example those in southern Benin and in Zaire (see page 38) and state farms in Ghana, most of

68 Western Africa

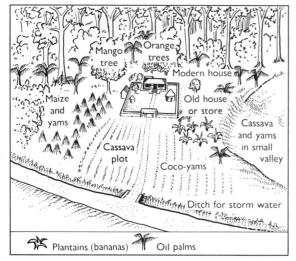

Figure 7.10 An Ibo mixed farm in the Niger Delta. Oil palms are mixed with food crops

the production in Nigeria is from small farmers. The Ministry of Agriculture tries to improve the quality of the product through agricultural extension services. It encourages the cutting out of diseased or old palms; and offers good seedlings and saplings to local schools for their demonstration plots. As in Zaire one of the problems is how to extract oil efficiently, and dozens of oil mills have been built throughout the palm-oil region. (See page 38 for the fact box on palm products.)

Cocoa in West Africa

In the 1960s cocoa was the most important single product exported from west Africa. It is now the second most important export after mineral oil.

There are four important cocoa-exporting countries: Ivory Coast, Ghana, Nigeria and Cameroon. Ghana once produced nearly two-thirds of the world total; now Ivory Coast produces 16 per cent, Ghana 8 per cent and Nigeria 8 per cent.

Cocoa farming in southern Ghana

This area is notable because it has enterprising African farmers and some of the best cocoa soils in the world. Find the Akwapim ridge on Figure 7.11. It was here that commercial cocoa growing first started.

Cocoa is grown in several different farm systems. This is because the new crop of the 1890s was fitted into the village crop system in the way that best suited it. This has resulted in

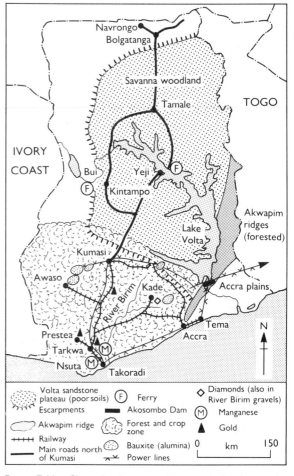

Figure 7.11 Ghana: landscapes and economic development

2 or 3 quite different patterns of land use. Look at Figures 7.12A and B.
1 Study the *strip-shaped cocoa and food farms* (Figure 7.12A).
The strip farms are notable because the farmers are migrants from farther east.
The men who formed the company to work the strip farms were not usually related. Each holding was measured on a road frontage and stretched back into the bush. It could be divided crossways or longways and given to sons. Farmers also bought other 'lands' further on, so that they could be worked by their sons and later left to them.
2 *Cocoa villages further west in Ashanti*
In Ashanti and other western areas zones of farming encircle the villages. The food farms are near the homes, the older cocoa land comes next, and the new cocoa occupies the third more distant zone (see Figure 7.12B).

The Guinea lands: coast and forest landscapes 69

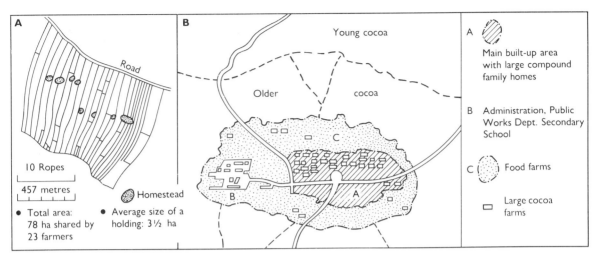

Figure 7.12A Strip shaped cocoa and food farms
Figure 7.12B An Ashanti cocoa village showing concentric land use zones round the settlement

The average holding of an individual farmer is about 2½ hectares. Nevertheless there are some large farms of 10–40 hectares, and even more.

3 *A family company of farmers*
The people of the Akwapim town of Aburi bought their distant farms on a family basis. The family organisation is quite different from the first group because inheritance is through the women's side of the family. Land is not owned outright, but is 'given' to members of the clan during their lifetime or for their use for a period. Hence land bought belongs to the *family* and is broken up in an irregular way according to need with the women's relatives getting a larger share.

There are many women landowners in all three systems; indeed in the western area there are as many women as men farmers. They were originally food farmers who planted cocoa and entered commerce. Some inherited cocoa farms from their mothers so the tradition of women growing cocoa is well established. Women play a prominent part in family and village life and trade in West Africa.

Marketing the beans

In Ghana the Cocoa Marketing Company now acts as a cocoa buying agent; in Nigeria there are several licensed buying agents. Both countries

Figure 7.13 This dark photograph shows the gloom of the cocoa forest in Ghana. The large leaves in the foreground are cocoyams. Both people farm. The man has a portable spray for insecticide. His wife carries a very large cutting *panga*

Facts: Cocoa

- Cocoa is a native tree of Central and South America.
- It was first introduced to the islands in the Gulf of Guinea.
- It was grown experimentally on the mainland at the end of the nineteenth century.
- Quaker families Cadbury and Fry offered a market to African farmers.

Cocoa needs:
- Temperature: 20–35°C
- Rainfall: 1,500–2,500 mm per annum well distributed; high humidity
- Altitude: up to 1,000 m
- Soils: well-drained, deep, loose-textured soils with much humus

Needs shade from sun in early stages and protection from winds.

Cultivation
Grown mainly on small farms (some estates Sao Tome, Liberia). Cocoa pods form on cacao trees which grow from 6–9 m high. Seeds are planted in nursery beds and trees transplanted. Mature after 5 years, maximum yield after 10 years. Two crops are harvested each year, main crop September–January and mid-season crop April–May, 30–40 pods per tree, each pod yielding 20–40 beans. Cocoa trees become the under-storey in forest. Mature trees have higher yields if they are unshaded.

Processing
- At farms and villages: ripe pods are picked, cut open and beans are removed by hand. Beans are fermented and dried in the sun, then bagged, and transported to collecting station.
- At buying centre: a random sample of beans is split and inspected to grade the beans (grade 1 cocoa must have less than 10 per cent damaged beans).

Production and exports
West Africa produced 57 per cent and exported 60 per cent of the world's cocoa in 1984. Ivory Coast, Ghana, Nigeria and Cameroon are important producers.

have marketing boards which fix the price to be paid to the farmer, arrange for the collection of cocoa, its transfer to a port, and the sale of the whole crop on the world's market. In good years some of the profits are set aside in a special fund to cushion the effect of a bad year or of a fall in world cocoa prices.

The hazards of cocoa farming: how research helps

Cocoa farming is beset by hazards. But agro-scientists have found remedies for diseases and ways of increasing yields. Three diseases are especially important:
- The *swollen shoot* virus attacked the crop in the 1920s and 30s when the only remedy was the cutting out of all infected trees. Then the mealy bug was identified as the carrier and controlled.
- *Capsids* (sucking insects) destroy young tissue and cause 'die back'.
- *Black pod* is a fungal disease.

Research also showed that mature trees yield better without shade and get most benefit from fertiliser in sunlight.

The hazards and details of cocoa farming have been spelled out because all major crops face similar hazards. Governments have to face the expense of scientific research if they want to keep domestic and commercial crops healthy and producing to capacity. An example is WACRI, the West African Cocoa Research Institute with laboratories in Ghana and Nigeria which share their research findings.

> 'Research is not much help unless farmers understand how to put it to good use.'

Even when solutions are found, there is an important task for the country's agricultural officers: to carry the information and the means of control into the smallest farm unit in the depths of the countryside. Spreading 'know-how' and encouraging people to use better methods is as important as the knowledge itself.

The Guinea lands: coast and forest landscapes

Liberia
Capital: Monrovia, 209,000 (1978)

Exports $428 million
Imports $367 million
Visible trade balance $61 million surplus (1983)

Export commodities	%	Export partners	%
Iron ore & concentrates	70	West Germany	25
Rubber	19	USA	23
Timber	6	Italy	13
Coffee	5	France	10

- 75 per cent forest cover well suited to rubber production.
- The four main iron ore producing regions – are all linked to Monrovia or Buchanan by rail.
- Foreign investment is encouraged and invisible exports are gained from the registration of overseas shipping.
- Nearly one-third of the people live within 80 km of Monrovia, but most people are still dependent upon agriculture.

Cameroon
Capital: Yaounde, 650,000 (1984)

Exports $1,162 million
Imports $1,078 million
Visible trade balance $84 million deficit (1983)

Export commodities	%	Export partners	%
Cocoa	22	France	33
Petroleum	22	Netherlands	25
Coffee	21	USA	12
Aluminium	7	Italy	6
Cotton	4		
Timber	4		

- Many contrasting climates, and a wide range of crops:
1. Wet coastlands produce oil palm, bananas and rubber.
2. The plateaus are good cocoa country.
3. The 'mist forest' area is a little cooler and bananas and coffee grow.
4. Above 1,800 m cattle graze the grasslands
5. The central savanna region produces cotton, groundnuts and millets.
6. Lower rolling savanna draining to Lake Chad in the extreme north has cattle, cotton and rice.
- There is an oil refinery, and also an aluminium smelter, using hydroelectric power generated at the Edea Falls and bauxite imported from Guinea.
- The Trans-Cameroonian railway provides a vital transport link.

Ivory Coast
Capital: Abidjan, 950,000 (1976)

Exports $2,591 million
Imports $1,314 million
Visible trade balance $1,277 million surplus (1984)

Export commodities	%	Export partners	%
Cocoa	20	France	19
Coffee	20	USA	12
Timber	13	Netherlands	12
Refined petroleum	9	Italy	9
Cotton	4	UK	4

- A well diversified economy based mainly on agricultural products.
- Coffee production has declined.
- Now the world's top cocoa producer (30 per cent of the world total).
- Since 1980, an oil producer.

Sierra Leone
Capital: Freetown, 470,000 (1985)

Exports $107 million
Imports $133 million
Visible trade balance $26 million deficit (1983)

Export commodities	%	Export partners	%
Diamonds	34	Netherlands	16
Rutile	16	UK	11
Cocoa	15	USA	9
Bauxite	13	West Germany	2
Coffee	9		
Palm kernels	4		

- Freetown takes its name from the colony of liberated slaves that was set up on the coast of Sierra Leone in the early nineteenth century, similar to Liberia.

Guinea
Capital: Conakry, 500,000 (1972)

Exports $538 million
Imports $236 million
Visible trade balance $302 million surplus (1980)

Export commodities	%
Bauxite & alumina	97
Pulses & oilseeds	3

- The second largest world producer of bauxite (15 per cent of world production).
- Deposits of iron ore and diamonds.

Chapter 8 Savanna landscapes in West Africa: the Sahel

> **Key words**
>
> Seasonal farming, erratic rainfall, inland delta, multi-purpose dams.

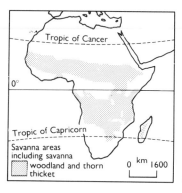

Figure 8.1 Savanna areas in Africa

This area of Africa seems to have had more than its share of disasters during the last 20 years. Hazards, such as drought and locust swarms, have been made worse for ordinary people by political actions including wars in Chad, Somalia, Ethiopia and the Sudan.

The term *Sahel* was the name originally given to savanna areas in West Africa. It is now used to describe drought areas right across Africa. The wet–dry savanna lands of West Africa form a broad zone 800 km from north to south and 3,000 km from west to east. They continue beyond the boundaries of West Africa as far east as the mountains of Ethiopia for over 5,000 km. There is a similar wet–dry savanna zone south of the equator and this means that the savannas are the most widespread regional type in Africa (Figure 8.1).

The dominant climatic feature of this vast area is the division of the year into two seasons, one wet and one dry. While it can be well-watered during the rains it also suffers acute drought, with 5, 6 or 7 dry months every year.

Figure 7.1 shows a division of the West African savannas into a southern and a northern section. The southern area has higher rainfall totals and a shorter dry season, and is better described as bush savanna, or savanna woodland. Further north, where there is a very long dry season and rainfall totals are less, there is more thorn scrub. In places there are huge baobab trees and some open grassland. The term 'tropical grasslands' is misleading. The savanna gives way to the Sahara Desert north of the Great Bend of the River Niger.

Look back at Figure 7.2 and note the climatic details relating to the savannas.
The savanna studies include:
- rural life in northern Ghana
- problems of the long dry season
- contrasts in water use along the River Niger
- seasonal changes in river level and how it affects everyday life
- the Inland Delta: a 1940s irrigation project in Mali
- Mopti, a river town; how the economy changes with the seasons
- Kainji, a modern multi-purpose dam in Nigeria
- the future of the Sahel: have people made the desert move south?

Rural life in northern Ghana

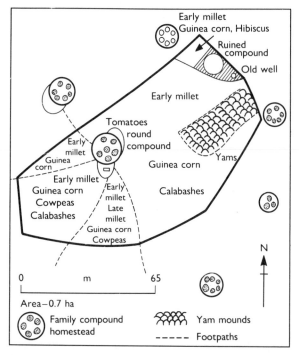

Figure 8.2 A small compound farm in northern Ghana

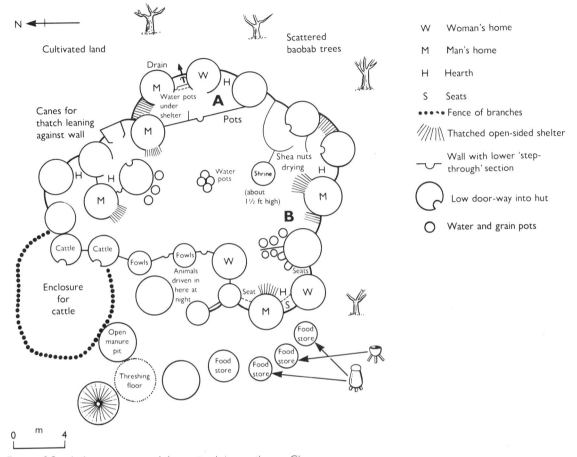

Figure 8.3 A large compound homestead in northern Ghana

The farm systems of the Sahel have survived because they are well adapted to local conditions. But if the rains fail, or there are more people to feed, or men leave the farm for wages work, they are put at risk.

Figure 8.2 shows part of a farm and homestead in northern Ghana. Farms average between 1 and 2 ha in size. Many have plots in the bush-fallow area where yams and other crops are grown. There is crop rotation and bush fallowing but homesteads are not often moved. Northern Ghana is fortunate in being fairly free from tsetse flies so that cattle are kept. Fields near the homestead are manured by cattle and household refuse. it has been estimated that in this area between two and five *tons* of a mixture of dried dung and refuse are applied to the ground of one family each year. A ton of dung is equal to about 20 kilos of superphosphate, and the yield of millet and sorghum is so improved that many 'near fields' produce crops of 150 kilos a hectare while bush farms produce 70 kilos a hectare.

Some farms have a lower-lying area where rice can be grown in between ridges which carry sweet potatoes or other root crops. Even path borders carry beans or nuts or flavouring plants which add relish to the meals which are mainly vegetarian, with occasional meat and eggs. This is a very intensive farm system. Where pressure on the land reaches danger point it is only kept going by the money sent or brought back by the men who go to work in southern Ghana. This area contrasts with the bush area further south where there are only 2 people to a square kilometre in the 'Middle Belt' (See Figure 9.10).

An extended family compound homestead in northern Ghana

Figure 8.3 shows a plan of the homestead of the Amnan family near Siniebaga. The diagram gives a picture of an extended family of more than 20 people. Each round house or *rondavel* is a good size, at least 3 m in diameter. There may be a raised bed or sitting place and a pile of soft hides to one side. The walls are often smoothed and highly polished. Sometimes there are shelves, pegs, and a clothes line over which things can be thrown.

Each junior family has its own walled-off section and its own hearth and bath place. A drainage gully takes the water under the main compound wall and out into the fields. Such a separate home is shown at point A on the plan.

The photograph, Figure 8.4, shows a view inside the compound, taken from B looking across to A. Shea butter nuts gathered from the savanna woodland trees lie drying in the sun. Immediately to the left is a shrine, possibly on the grave of a former member of the family. Walls are decorated in spare time during the dry season; huge scorpions are painted on some of the walls. The very large pots store water and grain.

The problem of the long dry season

All governments in dry-zone countries spend a high proportion of their budget on surveying, finding and bringing supplies of water to villages. Much has been, and is being done in Africa to improve rural life and reduce the drudgery of the long walk twice a day to get water, partly through Community Development Services.

Community Development runs literacy campaigns and classes, surveys needs, engages in village welfare, runs training courses, and lends money for building. In hundreds of places in different parts of Africa change is coming into remote areas.

Children as well as adults in Mali are building terraces to conserve water by making low rock walls along the contour. Weeds collect between the stones and hold back water and soil. Soil erosion is less and damper, deeper soil increases crop yields (see also page 178).

This work is rarely given the kind of publicity that Kariba or Kainji Dam receives. Yet it is equally valuable in the everyday life of village people. Some of the big projects bring electricity but African families must earn enough money to pay for it. Electricity helps small workshops and businesses and improves the quality of life.

Figure 8.4 A view inside the compound homestead. Note the shrine, the shea butter nuts drying, a pounding stone, large storage pots, etc.

Contrasts in water use along the River Niger

Seasonal changes in river level and how this affects everyday life

The River Niger is not only a resource but a unifying feature of the geography of West Africa. Most of the course of this great river is shared by 3 countries, and the headstreams rise in Guinea.

The Niger flood

The River Niger is about 4,000 km long. The course of the river divides into five sections shown on the map, Figure 9.7 on page 89. Starting at the source in the Futa Jallon and Guinea highlands these are:

1   The *mountain* section, where heavy rains (about 1,500 mm a year) fall from March to June when the monsoon thunderstorm zone moves inland from the coast as explained in Chapter 2.

2   The long *middle* sections of the river in Mali and Niger – about 2,000 km – have few tributaries. It is navigable for about 1,700 km for large river boats, except near the rapids east of Bamako. At Mopti the water is at its lowest at the end of the dry season in April.

3   Within the middle section of the river in central Mali, the river divides into a maze of channels and marshes, to form an *inland delta*. Flood water takes 3–4 months to pass through this part of the river. This low-lying basin was once an area of inland drainage like Lake Chad. In a wet period about 20,000 years ago the lake overflowed near Gao and made a new course to the Gulf of Guinea through what is now Niger and Nigeria. The inland delta fans the water out through a maze of natural and man-made channels.

4   By January the reduced flood passes through the rocky narrows of the Great Bend below the inland delta. It still takes several more weeks to pass across Mali and Niger to the *Nigerian* section and the reservoir formed by the Kainji Dam. The dam holds back over half of the peak flood so that water can be let out when needed during the period of low water. Fortunately the River Benue and its tributaries bring down a lot more water from the mountains of Cameroon. This doubles the flow of the Niger and maintains a seasonal flood and silt supply.

5   In the *Lowest* section the Niger again divides into several channels and forms a delta. The climatic figures for Warri, page 21, show that the area is sub-equatorial, with high rainfall and temperature all the year. It is heavily forested.

Along many parts of the Niger, this annual pattern of high and low water allows crops to be grown on the land that is seasonally flooded, or on damp soils after the flood has passed. Two areas where this is particularly important are in

Figure 8.5 A rice growing project near Timbuctu, Mali: planting floodplain rice. Settlements in the background are placed on the slightly higher levees or embankments

the inland delta in Mali, and in the lower reaches between the Benue and the main delta at the mouth of the river. Three examples show the different ways people along the River Niger adjust their lives to the water situation.

A 1940s irrigation project in the Niger inland delta

The inland delta is important for both farming and fishing:
- 250,000 ha can be flooded or irrigated for the production of cotton, rice, and other food crops.
- After the floodwater has passed, there is grazing for about 200,000 head of cattle.
- There are about 20,000 km of water where 50,000 fishermen catch about 90,000 tonnes of fish a year.

The inland delta has two main parts: the 'dead' delta below Segou and the 'live' delta, further downstream. The 'dead' delta is an area of gently sloping land which is no longer flooded naturally during the high water period. A barrage (a dam made up of movable sluice gates) was built at Sansanding in 1947. This raises the river level by a few metres, which is enough to feed the water down two distribution canals, the Canal du Sahel and the Canal du Macina.

The original intention was to irrigate a million hectares of the dead delta to grow cotton for export. However, the barrage did not hold back a lot of water, so that the peak passed quite quickly and there was not enough water to irrigate such a large area. About 40,000 ha of irrigated land has been brought into use, and over 30,000 people resettled. The emphasis has switched from producing cotton for export to growing rice for consumption in Mali (see Figure 8.5).

In the 'live' delta further downstream, natural flooding occurs. Rice growing is traditional here, and the area of cultivation has been extended by small-scale improvements such as embankments. There are major new irrigation schemes for rice growing near Mopti and Segou.

Mopti, a river town: how the economy changes with the seasons

Mopti is situated at the junction of the River Niger and its tributary, the Bani. The town is built on three low knolls within the flood plain of the rivers (see Figure 8.6). As the river water rises in May and June, the plains around Mopti

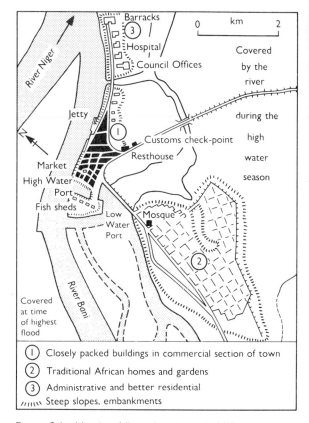

Figure 8.6 Mopti: a Niger river town in Mali

are flooded, leaving the town on its three islands linked together by bridges, and to the 'mainland' by an 8 km causeway. The rising water spreads over the land to cover hectares of rice seedlings. By December, when the water has begun to fall again, the rice crop can be harvested directly into canoes.

As the water falls, large numbers of cattle are brought in from the neighbouring hill areas for a few weeks for fattening on the rich grasslands. Then they start the long trek southwards to Ghana or the Ivory Coast.

The low-water period from March to May is the peak of the fishing season, for then the fish are confined to narrow water channels. Fish from a wide area are gathered at Mopti for export after drying or smoking (Figure 8.7). This is mainly in the 6 low-water months, February to July.

More than 90 per cent of the fish leaving Mopti is destined for foreign countries and is carried by lorry.

Savanna landscapes in West Africa: the Sahel

Figure 8.7 Dried fish at Mopti being packed for sale and transport. Note the woven baskets and matting, some used for lining crates. Cold storage and refrigerated trucks allow fresh fish to be delivered daily to Bamako and other places wherever the roads are passable, for distances up to 700 kms

Over half the fish exports go to Ghana, usually Kumasi, passing through Burkina Faso on the way. About 30 per cent go to the Ivory Coast and small amounts to Guinea and Burkina Faso.

Rice growing, cattle fattening, and fishing together form a varied but integrated economy with a distinct seasonal pattern related to the variations of the river that makes them possible. Thus, despite restrictions of site, Mopti has grown into a thriving town of over 15,000 inhabitants, the largest for 500 km.

Kainji: a modern multi-purpose dam in Nigeria

Kainji is a very good example of a large multi-purpose scheme. It is part of the Niger Dams project. It gets the benefit of two 'floods', one from the Guinea Highlands where it rains from May to October, but does not reach Kainji until March; and the second from local rains from June to October. These are enough to keep water flowing through the turbines and a good discharge below the dams.

It was intended to produce electricity, promote fisheries, navigation, irrigation and tourism, and enhance the Middle Belt (Abuja lies 300 km to the east). Changes in water saturation along the banks has resulted in vegetation changes. The increased woodland encourages tsetse flies (see Part 3, page 172) and the lake makes habitats for water snails that cause bilharzia and malaria mosquitoes.

Study the fact box on page 78 and Figure 8.8.

So how successful is Kainji? Together, Kainji and Jebba (100 km downstream) can cope with 'lower-than-usual' rainfall periods, both for electricity and irrigation. This must benefit development in Nigeria. But local people would have benefited more if:

- Lake fish catches had stayed at early high levels.
- A good rural water supply had been provided. No water taps were built for the new villages, so time is still wasted walking for water. Not even the new towns of New Bussa and Yelwa have adequate water.
- Improved water and sanitation would have been economically valuable and improved health.
- Control measures could have been planned similar to those at the Aswan High Dam. They include removing weeds that provide food for bilharzia snails and other measures to reduce infection.

78 Western Africa

Facts: Kainji Dam

- Started 1964: completed 1968.
- Installed generating power: 780 MW. Further power turbines not to be installed so that water can be released to suit the needs of agriculture rather than power production.
- Lake: 1,200 sq km, flooded over 200 villages (50,000 people) and 15,000 ha of farmland.
- Water control has reduced the peak of the flood by 60 per cent and has tripled the low-water flow.
- Some new farming takes place on the banks of the reservoir which are seasonally flooded. Most of the new farmland serving the new villages is not so productive because it receives less water than the old floodplain land. Below the dam there is now less seasonal fluctuation in water levels, so that less farming can take place on the floodplain. The reduction is up to about 50 per cent in the worst affected places.
- Fishing in the new lake was initially productive because the water was rich in nutrients from the flooded vegetation but soon declined. Below the dam, fish catches have reduced.
- The dam has reduced river blindness because the fly breeding sites have been eliminated both upstream (flooded) and downstream (less fluctuation).

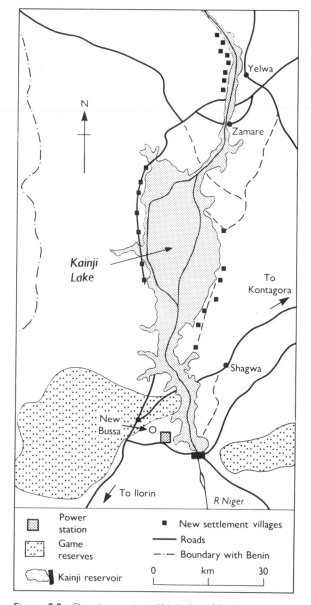

Figure 8.8 Developments at Kainji dam, Nigeria

Make a study of another large project in Africa, preferably one in your own country (see page 205). Which of the advantages and disadvantages listed for Kainji operate in your example?
1 Is your example multi-purpose?
2 Were local people moved and resettled?
3 If so, how well have the new villages and farms worked?
4 Are there any new problems, for example, reservoirs filling up with river silt?
5 What improvements could be built into a future plan?

Drought in the Sahel

People living close to the River Niger are fortunate because there is always a reasonable amount of water in the river coming down from the mountains. Further from the river, the population is dependent on the local rainfall, which is often unreliable. The Sahel area (see Figure 7.1) is regularly affected by drought, and its rainfall has been decreasing over the last 20 years and becoming more erratic. About 7 million people are at risk in this area between Senegal and Chad, where the population is increasing by about 2.5

Savanna landscapes in West Africa: the Sahel

Figure 8.9 The well: the main source of water for many scattered homesteads in Burkina Faso and other parts of the Sahel

per cent every year and food production by only 1 per cent.
- The 10 years from 1958 were among the wettest recorded in some parts of Africa and the cropped area and cattle herds increased.
- Then came a severe drought in the period 1968–74, when the rainfall totalled only 80 per cent of the average for a 5-year period.
- This situation was repeated in some areas in 1983–5.

In these periods of drought crops failed, about 40 per cent of the cattle died, and soil erosion allowed the desert to advance southwards by about 10 km per year. Famine brought death to thousands of people. Traditional herdsmen who had lost their cattle faced a complete breakdown of their way of life. It was very difficult to bring in emergency supplies because of the great distance from the coast, and the limited capacity of the roads and the two railway lines that enter the Sahel. The governments and peoples of the Sahel countries have been making great efforts to combat these problems The key objectives are that projects should:
- be small in scale, not large
- provide food for local consumption, not export
- save and store water in small ponds and dams (see Figures 8.10 and 8.11)
- favour the rural areas, not the towns
- reward people with the results of their own efforts
- depend on community self-help, not just outside 'experts'
- promote small-scale cultivation

Examples of these in Niger are numerous small vegetable gardening projects watered by hand from boreholes and wells, tree planting schemes to provide firewood, and new village shops built by the community selling bicycles, paraffin lamps, hoes, seeds, pens, salt, soap and oil.

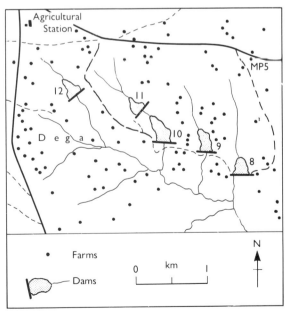

Figure 8.10 Settlement pattern and small dams near Bawku, north-east Ghana

80 Western Africa

Figure 8.11 The Sahel: cattle crossing a cement dam which holds back water and also acts as a 'bridge'

Mali

Capital: Bamako, 404,000 (1976)

Exports $177 million
Imports $255 million

Visible trade balance
$78 million deficit
(1984)

Export commodities	%	Export partners	%
Cotton & cotton products	54	France	14
Live animals	15	Ivory Coast	11
Groundnuts	13	West Germany	7
Fish	2	China	7

Niger

Capital: Niamey, 360,000 (1981)

Exports $576 million
Imports $878 million

Visible trade balance
$302 million deficit
(1980)

Export commodities	%	Export partners	%
Uranium	79	France	36
Live animals	12	Japan	18
		Nigeria	17
		Libya	15

Burkina Faso

Capital: Ouagadougou, 248,000 (1980)

Exports $126 million
Imports $360 million

Visible trade balance
$234 million deficit
(1982)

Export commodities	%	Export partners	%
Cotton	42	Ivory Coast	21
Karite nuts & oil	17	France	15
Live animals	13	West Germany	7

Chad

Capital: N'Djamena, 402,000 (1979)

Exports $78 million
Imports $99 million

Visible trade balance
$21 million deficit
(1983)

Export commodities	%	Export partners	%
Cotton	91	France	92
Live animals	1	Cameroon	3
		Nigeria	3

Mali, Burkina Faso, Niger, Chad

- These 4 countries share the following problems: inland position, poor and unreliable rainfall, frequent trade deficits, and a low per capita income. In all of them most of the people are dependent on subsistence agriculture, and on the marketing of cotton, groundnuts, and live animals.
- Cotton is the chief export of three of the countries, while uranium is the main export of Niger. But the world price of cotton and uranium has collapsed. This collapse has reduced per capita income in Niger by 30 per cent.
- Chad and Mali have valuable lake and river fisheries, and Mali has the vast Office du Niger rice and sugar cane irrigation scheme near Segou.
- The disastrous droughts in the Sahel caused great hardship. Harvests failed, livestock herds were destroyed, people starved. The rainfall situation has improved, and many small water storage and irrigation schemes are being carried out to try to avoid a repetition of the problems.
- All 4 countries are among the poorest in Africa, and are heavily dependent on foreign aid.

Senegal

Capital: Dakar, 850,000 (1979)

Exports $481 million
Imports $973 million

Visible trade balance $492 million deficit (1983)

Export commodities	%	Export partners	%
Refined petroleum	19	France	32
Groundnuts	17	Mali	10
Phosphates	16	Ivory Coast	8
Fish	14	Mauritania	6
Shellfish	6	UK	6
Sea salt	3	Guinea	3

- Well diversified economy.
- Important world producer of both groundnuts and phosphates.
- Petroleum products from imported crude oil made at the refinery and petrochemical plant near Dakar are exported to neighbouring countries.
- Dakar continues to be an international business centre for Francophone West Africa, and has extensive industrial and distribution activity.

The Gambia

Capital: Banjul, 49,000 (1980)

Exports $45 million
Imports $129 million

Visible trade balance $84 million deficit (1981)

Export commodities	%	Export partners	%
Shelled groundnuts	55	Netherlands	24
Groundnut oil	23	UK	18
Fish	11	Italy	16
Groundnut meal & cake	8	Belgium/Lux	7

- English-speaking state surrounded by French-speaking Senegal: river waterway cut off from hinterland. Cooperation is essential, if dams are to be built on the upper part of the river for electricity and irrigation.
- Confederation of Senegambia formally declared in 1982, confederal cabinet and parliament met in 1983 for the first time. Progress towards full union is very slow, perhaps wisely.
- Groundnut production has declined. Aid projects have been focused on rice production for export and to save foreign exchange.
- The problem of introducing new crops, or new methods, or more cash crops, is that it disrupts traditional practice. In the Gambia rice was one of the food crops grown by women. Their problem was always to get enough help with the heavy work because men were more concerned with groundnut farming for export, because that was where the money was. Increased rice exports have resulted in a drop in domestic food supplies. The women still do the heavy transplanting of rice seedlings and weeding.
- Coastal and river fishing is being improved.
- Tourist trade is being developed.

Chapter 9 Urban life and development in West Africa

> **Key words**
>
> 'Models', core area, multi-functional, infrastructure, federation, energy network

Urban life has always been strong in West Africa. There is a higher proportion of people living in towns than there is in the countries of east and central Africa. In West Africa there are over 30 cities with over 100,000 people, as much as 12 per cent of the population. In other parts of Africa 20,000 people counts as a good-sized town.

Growth of West African towns

Town	1930s	1950s	1980s
Abidjan (Ivory Coast)	18,000	119,000	1,250,000
Accra (Ghana)	71,000	117,000	1,100,000
Bamako (Mali)	21,000	101,000	450,000
Dakar (Senegal)	93,000	251,000	980,000
Kano (Nigeria)	89,000	130,000	750,000
Lagos (Nigeria)	137,000	267,000	5,000,000
Ibadan (Nigeria)	387,000	459,000	2,000,000

The agricultural town in West Africa

West African towns often cover a considerable area and are protected by walls or ramparts within which there is also farmland. Within the old walls of these cities some homes are built closely together, but others have their own garden surrounded by a high wall.

Whatever their original size, West African towns are not only increasing in population and area, but are taking on international characteristics of function and form as well as buildings similar to those found in many other parts of the world.

The growth of towns

As towns develop a similar pattern of growth can often be found. When the same pattern is repeated over and over again it becomes a 'model' of a *theoretical town*.

Several 'models' of towns and their functional zones are shown in Figure 9.1A and B. Zaria, Kano, Lagos, and many of the larger cities have separate functional zones, and are called polycentred. Zaria is shown as an example (Figure 9.1C).

Find and note:
1. *The old walled city.*
2. *An older Hausa capital* which covered a much larger area showing Zaria's long history.
3. *The township* – the colonial centre (1900s) which included a Government Reserve Area to the north of the river and the railway station (1911).

Now:
4. Write down the different functions of the area within the township.
5. *Tudun Wada* became a third nucleus for African workers, especially those who came from the north. What goes on there as well as places to live?
6. Name 3 other nuclei to the north-west beyond the river, that grew up later still.

The districts or neighbourhoods in West African towns are often based on kin or ethnic groups because family and kin ties are very strong. In Zaria, Sabon Gari (a township to the east of the railway) was originally the home of immigrant workers who were not Muslim and did not come under Islamic law. In southern Africa neighbourhoods are more often segregated by class, race or colour.

Outsiders are impressed by the mobility of African people: of men who go off to their farms, or to other countries to work; of the women who go to markets both as buyers and sellers; of children who carry water or collect firewood and have a long walk to school or students who 'travel' to seek education.

African towns seethe with people who come to find work and greater opportunities than their villages provide. Parts of African cities are really suburban villages where people have moved in and built the kind of house they knew at home. One of the problems for Africa is that village hygiene, which is satisfactory in the countryside,

Urban life and development in West Africa 83

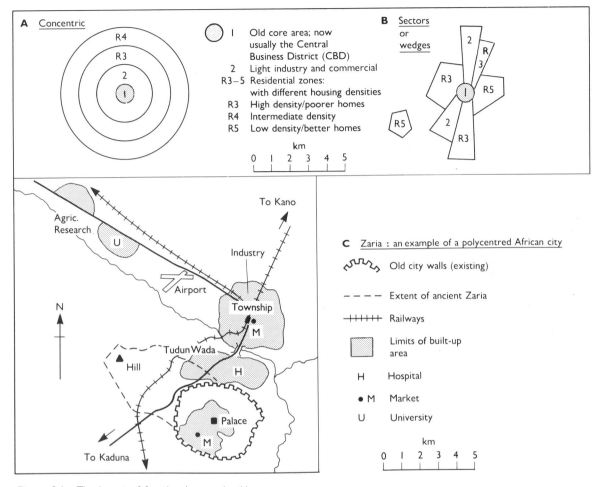

Figure 9.1 The layout of functional zones in cities

does not adapt well to the over-crowded conditions of a town, and it is a problem to know how to provide plumbing and refuse disposal in parts of town–villages where some censuses say that people are living at a density of over 700 people per sq km. If people leave their home area and travel to find work or schooling, it is natural that they should go to stay with someone from their home town.

Throughout Africa all that is needed to set up shop is a piece of ground and a shady tree. In town or village or market, barbers cut hair, tailors and seamstresses work their sewing machines, and the letter-writers tap their typewriters. All these are informal ways of making a living, see Part 3, page 192. In the markets it is a roof only that is needed, as a shelter against rain or sun, and shops and homes have verandahs.

Lagos: a multi-functional metropolis

1. Find the map extract box, page 96 before reading on. Figure M1, in the box, is a map showing the site and main functional areas of Lagos. The city centre and main port area is shown in more detail on map extract, Figure M2, page 97.
2. Check that you understand the meaning of *site, situation,* and *function* (Mombasa, page 58).

Origins and development of Lagos

The original settlement grew up on Lagos Island which is a little over 3 km long and less than 2 km wide. Although small, it was a trading base during the nineteenth century for 3 groups of

84 Western Africa

Figure 9.2 The city of Lagos. Note: contrasts in modern building styles, both business and residential; part of a creek in the centre; the position of an elevated road or 'freeway'

people:
• The Yorubas, occupying the old town in the western third, whose great cities were inland.
• The 'Brazilians' on the northern side who were Africans freed and returned from slavery in Brazil. They were often wealthy and had 'know-how'.
• The colonial traders on the southern side of the island.

This island is still the core of the city. Some of its contrasts are shown in the photograph in Figure 9.2.

Development has spread outwards from Lagos Island, particularly along the railway line and main road northwards. Bridges have been built to link the four main islands and the mainland. It is now about 30 km from the coast to the northernmost suburbs.

As in Mombasa, different parts of the city have developed into specialised functional areas. The central business area (sometimes called CBD for central business district) extends along the south-west side of Lagos Island, with many important office buildings and public facilities. The main industrial and port area lies on the west side of the harbour at Apapa (see Figure 9.3).

Apapa Trading Estate is an example of an early industrial area where access roads, rail and power were provided before both small and large firms rented sites and built factories. There are more industrial areas in the northern part of Lagos near the airport. However most of the built-up area is used for housing the population of possibly 5–6 million people. The residential areas vary from the closely packed houses of the old town to the spacious villas on Ikoyi and Victoria islands.

Big city functions

Lagos is now one of the principal cities of tropical Africa. It is a multi-functional city: a port, an industrial centre, and a focus for administration at regional, national, and international level.

Most large cities are important for business activities. These are the places where:
• commodities are bought and sold
• arrangements are made for goods to be sent out to shops and factories in other towns
• the administration of government affairs takes place

These activities in turn need support services such as airports and taxis, hotels and restaurants, telephone and telex services, and agencies to provide secretarial, computing, and photocopying services. This is where important people meet and important decisions are made. Thus it is often a good place to print newspapers and publish magazines, and to establish colleges and universities.

All these are especially important in capital cities for these are where governments meet and foreign embassies are found. Hence large cities have large numbers of office buildings, usually clustered in the central business district (CBD) where it is easier to bring people together for meetings and to attract workers. Lagos is no exception, and large office blocks for commercial firms and government departments cluster on Lagos Island. There are also offices for state departments, and international agencies such as the following:

Organisation	Full title
ECOWAS	Economic Community of West African States
ILO	International Labour Organisation
UNICEF	United Nations International Children's Emergency Fund
UNDP	United Nations Development Programme
WHO	World Health Organisation
UNESCO	United Nations Educational, Scientific, and Cultural Organisation
CDC	Commonwealth Development Corporation

There are offices for West Africa Health Council, Port Management Association of West Africa, Cocoa Producers Alliance, World Bank, Amnesty International and many others.

The need for a new capital city: Abuja

There are two main reasons why there is a new federal capital for Nigeria at Abuja.
1 First is a political reason: the government's desire to knit together the separate states of the country into a working whole.
2 Second is concern about the future of Lagos if drastic action is not taken:
- *water supply*: the demand for water was already higher in 1976 than the amount forecast for the year 2000
- *sewage*: a report said 'only a small part of the city has any semblance of a modern sewage system'. Providing fresh safe water, and getting rid of waste water and sewage are both hindered by the near sea-level altitude of Lagos
- *housing*: there are 300,000 people living in improvised homes. One estimate says that 2,500 people move into Lagos every month, and that one-third of the housing is unfit. The provision of essential services can only be achieved by harsh clearance of homes in some places and a master drainage plan.

Lagos is not a simple city on the European model.
- The extended family, village ties and tribal

Figure 9.3 Apapa Quays, Lagos, looking southeast across the lagoon towards the sea. The quays and transit sheds have been built on reclaimed 'flats' like those in the middle distance on the right. The most distant line marks the coast and breakwater

loyalties persist. The suburbs are clusters of family and tribal groups.
• The land tenure system of the countryside often operates: the *family* rather than an individual owns property. Finding the many owners of land and persuading them to sell, in order to build new major roads or provide piped water and sewage disposal is a major headache.

The explosive growth of Lagos, with many different functions on a difficult island site, has caused great problems in providing the supporting *infrastructure*, that is, water and power, drainage, transport and telecommunications, schools, and other services. Despite the massive new road system, there is serious traffic congestion, and problems with the other services. These features are repeated in the other great cities of Africa, such as Kinshasa, Nairobi, and Cairo.

The Nigerian government decided in 1975 to reduce the pressure on Lagos by moving as much of the national administration as possible to the new capital city at Abuja. The new Federal Capital Territory covers an area of 8,000 sq km near the centre of the country, where it is well placed to serve all regions, and to be a focal point for national unity.

The new city is one of the biggest construction projects in the world. The focal point is the city centre with adjoining zones for government buildings of all kinds. The residential areas can expand progressively along 4 development corridors, separated by open spaces and main roads. A great deal of effort has been put into the early stages of the city, and it is hoped to transfer more government departments from Lagos. The population may rise to 500,000 by the end of the century.

Lagos has been studied in detail because it shows how rapidly cities in Africa are changing. Despite its problems of siting, Lagos is a thriving, lively and rapidly expanding city of some 5–6 million people. When all the federal functions move to Abuja, Lagos will still be a busy major port and commercial, industrial and conference centre. Taking off some of the pressure will allow elbow room for these activities, improve efficiency and ease the daily life of a great city.

1 Use the information in map extract box, page 96, and the above text to summarise the chief features of Lagos, as though you were answering an examination question.

2 Use the same headings as for Mombasa, pages 58–59:
• *site*: island, type of coast, harbour advantages, disadvantages
• *situation*: in relation to the rest of Nigeria, West Africa and world communications
• *function*: port, capital city, industry, services

Mineral oil in the Niger delta

The Niger River delta to the east of Lagos is one of the most important areas in the world for mineral oil.

Study Figure 9.4 and note the following.
1 The Nigerian National Petroleum Corporation (NNPC) works with each of the foreign companies. The oil companies have a 40 per cent share of the profits. The remaining oil revenues and taxes benefit the government and the state in which oil is produced. In some years oil revenues provide 80 per cent of government income.
2 Use the key to find:
• the offshore exploration and producing areas with blocks worked by different companies
• the main producing oil fields (a few offshore, most inland), their limit shown by a bold line (exploration is still going on in this area)
• further inland are the 'open acres' where some concessions and service contracts can still be taken up

Drilling conditions in the hot humid swamp of the delta are some of the most difficult in the world. Teams work in mud and water, attacked by mosquitoes. (Check the climatic figures for Warri, page 21.) There are overcast skies and drenching rain and the forest drips grey and forbidding behind the bleached roots of the mangroves. The first discoveries and development were in the eastern Niger delta where pipelines carry the oil to Bonny terminal.

Natural gas is also being used to generate electricity and if this can be fed into a grid it will provide both industrial and domestic electricity in a countryside where kerosene lamps are still common. Thus oil and gas can provide immense opportunities in a country which is generally short of power and has only one notable coalfield in Enugu in Anambra State.

Nigeria's integrated energy plan

Figure 9.6 shows how different sources of energy

Urban life and development in West Africa 87

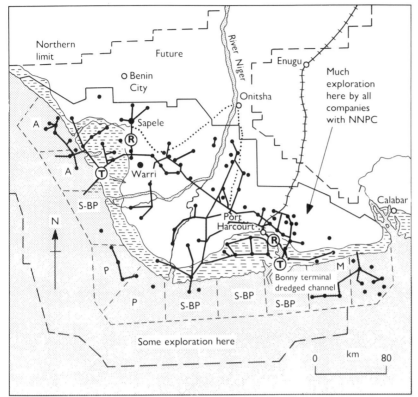

Figure 9.4 Niger delta oilfields. Concessions are shown only in the off-shore area but the whole delta is held by various oil companies, working with the Nigerian government

in Nigeria are being integrated into an energy network so that power can be extended to distant parts of the country.
Find:
- the main electrical power routes
- the petroleum/oil network of depots and pump stations
- the new gas gathering system to fill the proposed gas grid. In the 1960s natural gas had to be burned off as waste. In the 1990s it will be piped into the gas grid. Only the pipe to Port Harcourt operates at present.
- the hydroelectric station for the Kainji dam.
   The priorities for the 1990s are:
- integrated energy development
- replacement of all worn or out-dated machinery
- a 4th oil refinery (the 3rd is at Kaduna)
- development of petro-chemical industry
- complete gas gathering

The Federal Republic of Nigeria: a giant of West Africa

Nigeria is the giant of West Africa because of its very large population (92 million) and the dominant position of Nigerian oil (about 75 per cent) in the total value of exports from all the West African countries. Oil development has had a dramatic impact on the Nigerian economy, which was dependent on agricultural products for 60–80 per cent of exports in the 1960s.

During the years of the oil boom (1960–74) oil revenues were used to finance a vast range of improvements to infrastructure, including oil refineries, dams for electricity and irrigation, and a motorway system intended to help deal with traffic congestion in Lagos. Agricultural and community development projects were started.

But the worldwide recession has reduced the demand for oil. Oil production and oil income have both fallen in Nigeria so the government receives less money from oil royalties, profits, and taxes, and has had to cut back the various public investment programmes.

The 'fat years' of the oil boom have left problems.
- A large food import bill for wheat and other exotic foods costing 15–17 per cent of the value of imports in most years. Nigerians have changed their eating habits. Rice has now to be imported

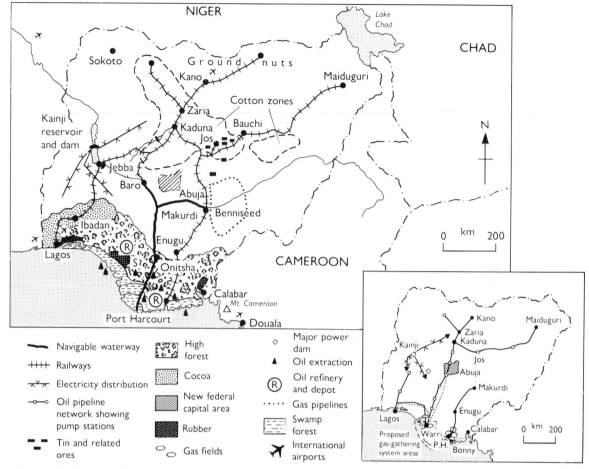

Figure 9.5 Economic development in Nigeria

Figure 9.6 Nigeria's national energy network

(though it can be grown in Nigeria). Formerly, the basic foods were millet, yams and cassava.
• In spite of capital spending on large agricultural and irrigation projects agricultural production only increased by 1 per cent. But Nigeria's population grows twice or even three times as fast. Agriculture could provide valuable exports if there was an increase in production of 5–7 per cent.

In spite of the decade or more of 'lean years' following the oil slump, Nigeria remains a very wealthy country with a varied range of resources, a large number of important rural development projects and a very well-developed industrial base (see Figure 9.5). The large population (more than all the other West African countries together) forms a good market for local products. The federation will continue to be supported if people believe that it will provide benefits for them. If the system works unfairly, the tribal or social groups that are at a disadvantage will want to change the system.

West African Cooperation: ECOWAS

The governments of many of the West African countries have recognised that they will have to work together if they are to overcome the problems caused by:
• the colonial pattern of national boundaries and transport links (see Figure 9.7)
• the drought

Urban life and development in West Africa

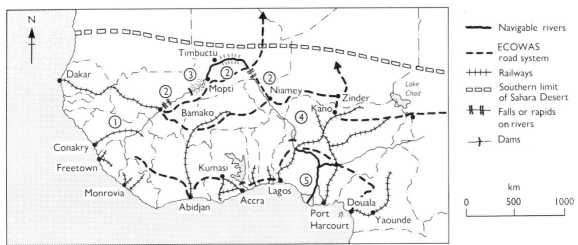

Figure 9.7 Western Africa: transport links. There are a large number of fairly small, well populated countries along the coast, and a few large countries with small populations in the interior. The European powers established their initial settlements on the coast, then pushed inland to find trade resources and markets for their goods. The artificially shaped countries such as Togo, Benin, and Gambia, and the railway lines constructed far into the interior are the results of this period. ECOWAS development plans aim at improving the poor west–east links to supplement existing south–north roads and railways

- the world market price system which has increased the costs of manufactured products much more sharply than the value of the raw materials and agricultural products exported by the African countries.

In 1975, 16 nations in West Africa formed ECOWAS – the Economic Community of West African States. The countries hoped to create a regional common market of over 150 million people where there would be no customs barriers to hinder trade, and free movement of capital and labour. It was also hoped that there would be a common currency and monetary zone, and a coordinated economic strategy to make the best use of the limited resources. Transport links between the countries would be improved by the construction of a west–east road in the coastal zone, and another roughly parallel, further inland.

Despite these high hopes, there has been little real progress. Most of the countries have not paid their membership dues, and few of the proposed projects have been started. This lack of progress is due to:
- A very difficult period of world recession. Most countries were too concerned with their own problems of debt repayments, unemployment, and drought to think about international cooperation.
- The countries in the same climatic zones produce similar crops which compete in overseas markets (see Figure 9.8).
- Each country wishes to develop its own local industry rather than import from a neighbour.

It is a sad fact that in West Africa today
- Linking up railways is difficult because lines were built on different widths
- West to east roads are tediously slow
- It can take up to 6 months to complete a banking transaction between two neighbouring countries
- the post and telephone system is inefficient

Is the concept of ECOWAS too ambitious to succeed over such a large area?

What might be possible?
It is important to:
- have small but achievable targets – 'don't run before you can walk'
- rely on a great deal of self-help
- focus on areas of development likely to benefit all countries – people will support what can be seen to work
- focus on a good road network, improve post and telephone services, reduce friction points
- continue to publish reports in French as well as English and teach both languages in schools

90 Western Africa

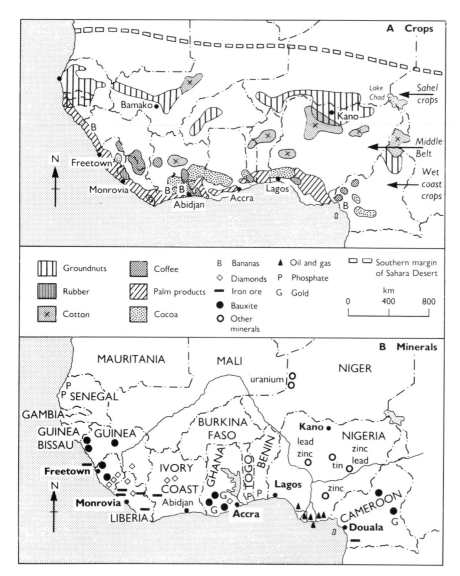

Figure 9.8 Western Africa: commercial crops and mineral resources

Ghana Capital: Accra, 965,000 (1984)

Exports $566 million
Imports $533 million
Visible trade balance $33 million surplus (1984)

Export commodities	%	Export partners	%
Cocoa	45	USSR	19
Gold	24	Netherlands	15
Timber	3	UK	15
Manganese	1	West Germany	9
Diamonds	1		
Bauxite	1		

Nigeria Capital: Lagos, 5 million (1980s)

Exports $11,882 million
Imports $8,900 million
Visible trade balance $2,982 million surplus (1984)

Export commodities	%	Export partners	%
Petroleum	99	USA	46
		Netherlands	12
		France	10
		West Germany	7
		UK	2

Northern Africa

Chapter 10 The Maghreb

> **Key words**
>
> Muslim world, contrasts in land use, alluvial fan cone, gravity flow, irrigation, water and sun as resources

North-western Africa is sometimes called Barbary after the Berbers, the chief inhabitants before the Mohammedan invasion in the eighth century AD. It is also called the Maghreb, a shortening of the Arab name *Djezira el Maghrib*, the Isle in the West. It suggests that the Maghreb countries, Morocco, Algeria and Tunisia, are a separate group – an island – between the Mediterranean Sea and the Sahara.

In other ways the Maghreb is not isolated. It is separated from 'Africa south of the Sahara' but it has close ties with Europe and Asia. Geographers sometimes ask whether Africa ends at the Pyrenees or Europe ends at the Sahara. Today it forms the westernmost limit of the Muslim world, but retains its close links with Europe.

This chapter is about the coastlands and mountain zones of the 3 Maghreb countries. But about four-fifths of Tunisia and a great deal of Algeria lie within the Sahara Desert. This area, and the mineral wealth that it produces, is covered in Chapter 11 on the arid lands.

The Maghreb is a land of contrasts. A journey across it includes towering snow-capped mountains, gorges, plateaus, salt-wastes, terraced slopes green with young vines, almond groves bright with blossom, dark forests, busy Arab cities and markets, sophisticated hotels and beaches, and everywhere irrigation water. Perhaps in the last of these is one of the main keys to development.

There are some ways in which the Maghreb resembles West Africa.
- Both have distinctive landscapes parallel to the coast.
- Second, the savanna of West Africa and much of the Maghreb are wet–dry lands, but with quite different types of wet–dry climate.
- From October to March the Maghreb experiences European weather, or rather, European rainfall with milder temperatures, more sun and a very early spring. From May to September it has Saharan weather when the weather belts move north; and it is very hot and very dry.
- The main difference between the two areas is that the savanna of West Africa has rain in the hot season, while north-west Africa has its rain during the northern winter months, and the summer is rainless.
- Because of this the Maghreb can grow the crops of two 'worlds': temperate crops during winter and spring, and irrigated subtropical crops during the very hot dry summer.

Contrasting landscapes

Figure 10.1 shows the contrasting landscape types – plateaus, mountains, deserts and coasts – on a sketch map. Figure 10.4 in the practical work box on page 93 summarises the changes in land use related to climate and relief from the Mediterranean coast to the Sahara.

In the north are the mountain ridges and plain called the *Tell*. Here the rainfall varies between 400 and 700 mm and there are 3 or 4 dry summer months.

In the central plateau area the rainfall is usually below 400 mm, but it is much heavier on the bordering ranges where there is snow in winter, lasting into spring or even early summer. Even on the plateau itself the autumn-sown grains profit from the slow melt. Potentially the plateau areas are excellent wheat and barley lands. There are broad open plateaus and grain silos or elevators along the railway, reminding one of the Canadian Prairies. This area was one of the granaries of the Roman Empire. However, the rainfall is sometimes unreliable and there are saline soils near the salt basins called *shotts*.

In the Atlas mountains bordering the central plateau higher rainfall totals and winter snow result in stands of coniferous forest. There are winter ski resorts and other mountain tourist centres, especially in Morocco.

On the south side of the plateau and the Saharan Atlas the rainfall is only 100 mm a year or less and the desert begins. However, rivers flowing south from the mountains give life to superb oases before their water is lost in desert sands.

92 Northern Africa

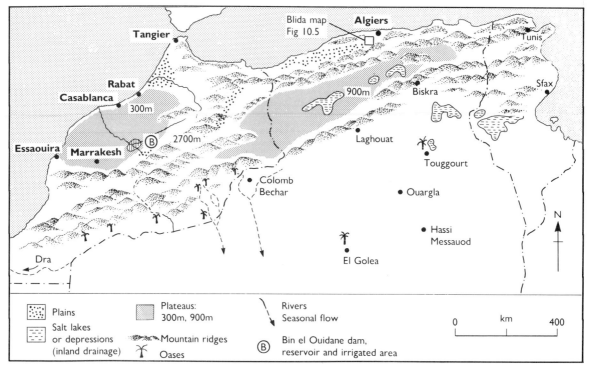

Figure 10.1 Landforms in North-west Africa

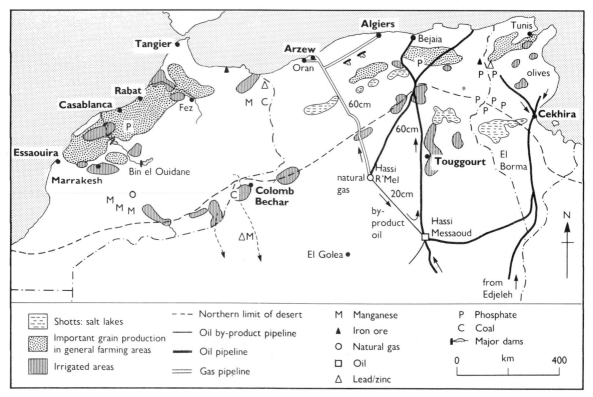

Figure 10.2 Economic development in North-west Africa

Practical work box 6:
how to use transects to organise information

1 A transect is different from a section:
- a transect shows a journey across the land surface
- a section cuts down through the earth to show what lies below the surface, for example, a gold reef in a gold mine (see Figure 17.5).

2 Transects are useful for summarising information:
- they show changes over a large area
- they can combine different kinds of information, for example, height of land, weather, vegetation, products

3 Look at Figure 10.1 and 10.4. Figure 10.4 is a transect across the map, Figure 10.1 to the west of Algiers. The scale of the transect is much larger than the map. The spot heights shown on the map are 900 m in the centre of the Algerian high plateau, and 300 m and 2,700 m in Morocco, the latter in the High Atlas mountains. Sea level is zero.

Work like this is to draw the transect

1 Copy the *frame* from Figure 10.3A onto your paper.
2 Use a strip of paper to measure positions along the base of the transect (see Figure 10.3B). Where there are two or more heights next to each other, put a sign to show if the land is above or below the height line.
3 Take your marked strip and plot the points onto the frame (see Figure 10.3A). Join them up to make the profile of the land surface.
4 When you have finished the land surface profile, label each landform *at the top of the frame*. It is helpful if you draw lightly dotted *vertical* lines down to separate main areas from each other (see Figure 10.4). Leave a column on the left side to write in headings.
5 Decide which information you need and write in the subject headings at the left-hand side. Use the text and diagrams in this chapter to fill in the columns.

Such a diagram is called a *table*, and the method is to *tabulate* in groups to show how bits of information relate to each other.

Figure 10.3 Frame and marked strip of paper

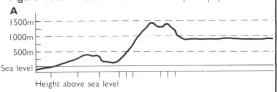

Figure 10.4 Transect north to south across the Maghreb and related table

Landforms	→		Mountains	Plateau		Mountains	Foothills, Oases	Desert
Place names	NORTH Algiers Blida							SOUTH Touggourt
Rainfall climate	Winter rain Hot summers					Winter snow		60 mm or less
Natural vegetation etc	Wooded hills	Forest		Grass thorn scrub		Forest		Desert
Economic products	Mediterranean fruit and vegetables. Tourism			Grains pasture (sheep)		Tourists	← Oasis → products	

94 Northern Africa

Land use and relief near Blida, Algeria

The site of Blida is shown on the transect, Figure 10.4, and on Figure 10.1 as a very small box. Figure 10.5 is an enlargement of the Blida area. It is included because it adds detail to that part of the transect where the Tell Atlas mountains meet the plain of the Tell. More important, it shows how rivers and streams build up cones and terraces in the 'foreland' zone bordering mountains.

These become important sites for settlement and farming. Silt laden rivers lose speed and drop their load at the junction of mountain and plain. This forms a large triangular alluvial fan, or if there are streams close together, a 'foreland' zone a little higher than the plain or plateau. Similar landforms occur worldwide, and lines of oases and many important towns in Africa are sited in places like these.

1 Check Figure 10.5 and the text to find:
- a mountain range in the south
- the Tell plain in the north
- the site of the town of Blida
2 Check the symbols for roads and the built up area. Blida stands out clearly and the roads fan out from the river gorge, showing the shape of the cone.

The slope of the land makes it easy to distribute irrigation water by natural gravity flow. It can be increased by pumping. Water is taken off by a canal to the south, slightly upstream. It then fans out through a mass of smaller channels to Les Orangeries. The exit points from the Blida built-up area are shown by arrows. Some water is even taken along gutters through the town.

Les Orangeries is a very intensively farmed area. It grows speciality market garden crops such as early vegetables and citrus fruit because it is high enough to escape the spring frosts of the plain.

The Tell plain. The 200 m contour line

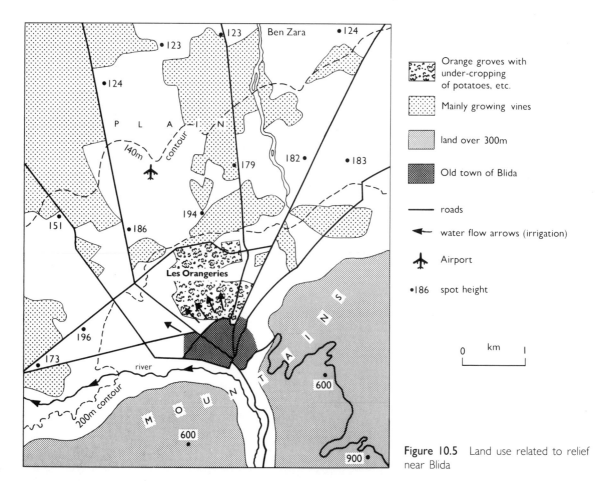

Figure 10.5 Land use related to relief near Blida

separates Les Orangeries on the higher ground from vineyards below that level. Vines can stand moderate winter frosts.

This small area illustrates some of the most important features of the geography of the Maghreb:
• three landscape types: mountain, plain and alluvial cone
• the varied use of contrasted areas
• the importance of water and its conservation and distribution, especially during the hot, dry, summer months

Water and climate as resources for development

The following pages describe:
• the development of water resources in north-western Africa to provide irrigation water
• climate as a resource
• cities and economic development

Figure 10.6 Irrigation and land use at Beni Moussa, Morocco. The dam at Bin el Ouidane stores water which is piped through the mountain. It provides a head of water for the turbines of the power house at Afourer (shown on photograph) and for irrigation

The development of water resources is a key element in the increase of agricultural production in the Maghreb and could contribute to a general improvement in the standard of living throughout the area. Fortunately the mountains receive a higher rainfall than the neighbouring plains. The winter snows regulate the flow, and make water that would otherwise be lost available during the dry summer months. An immense amount of money is being spent on water projects. There are many dams for water storage, with canals and channels to distribute water to irrigated crops. A few of the largest dams are marked on the economic development map (Figure 10.2). One of the most spectacular is Bin el Ouidane at Afourer in Morocco, serving thousands of hectares of irrigated land. The main distribution canal and some of the farmland is shown in the photograph, Figure 10.6.

The winter climate of the Maghreb also provides a valuable resource in two ways.
• Most of Europe beyond the Mediterranean shores to the north lacks sunshine and warmth during the winter. The countries around the Mediterranean have a very mild early spring which makes it possible to grow fruit and vegetables much earlier in the year than in northern Europe. Very high prices are paid in Europe for produce that is sold weeks ahead of the main European crop. The old raised beaches along the Atlantic and Mediterranean coasts of the Maghreb provide light, warm soils slightly above sea level. These are protected by cane fences from salt sea winds and frosts, and are excellent for early vegetable crops such as tomatoes, potatoes, carrots, and celery, which can be marketed from January onwards.
• The sunshine also attracts tourists, and north-west Africa can offer spectacular beaches and scenery, and historic Arab cities. Air travel allows people from northern Europe to escape to the sunshine in the winter. Tunisia has been particularly successful in creating holiday villages along the coast and on the islands. Up to about 2 million visitors come each year, and invisible income from tourism is the most important source of wealth after petroleum.

City life in northern Africa

The Maghreb and the north African coastlands have a long tradition of urban life. In some places there must have been almost continuous settlement, perhaps for 3,000 years or more: Phoenician, Carthaginian, Greek, Roman, Islamic, Turkish, European.

Most towns are large and sprawling with *souks*
continues on page 100

Map extract box 1: the port and city of Lagos

There are 4 survey map extracts in this book. The scale of three of them is the same, 1:50,000 but their subject matter is quite different.
- *Lagos*: a port and metropolitan city (Nigeria, 1964)
- *Mount Kenya*: (scale 1:25,000) physical geography of a glaciated highland area (Kenya)
- *Nkana-Kitwe*: a copper mining area (Zambia, 1986)
- *Makwiro*: contrasting land-use patterns (Zimbabwe, 1982)

Look at Figure M1 (1984) and compare it with the 1964 map:
- only one-tenth of the built-up area of Lagos appears on the map extract
- marshy ground and the lagoon site severely restrict the land which can be built on
- there has been an enormous expansion of the built-up area northwards in 20 years

Old maps can tell us about the amount of change and the kind of development taking place if they are compared with what we know of a place today. Now see what answers your detective work on the Lagos map extract gives. Use Figure M1 to help with your answers. Before you begin, check page 58 to make sure that you understand the terms site, situation and function.

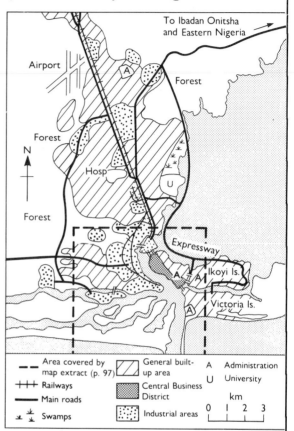

Figure M1   The site of Lagos

1. The site of Lagos
(a) Is the built-up area on the coast or not?
(b) What makes it difficult for the buildings to expand southwards?
(c) The main port development is at Apapa. Why has this advantages over Lagos Island? In which direction has the port and industrial area extended?
(d) How is the entry to the harbour protected?

2. Study the coast and the lagoon
(a) From evidence on the map, is the coastal area more likely to be rocky or mud and sand?
(b) What is the main direction of the coast? Would you describe the seaward coast (as opposed to the inner lagoon coast) as straight or irregular?

3. Communications
(a) What different means of travel and other communications are shown on the maps?
(b) How many major bridges have had to be built because of the island site of Lagos?
(c) Try to sum up the balance between the difficulties and advantages in communications.

4. Contrasts in land use
Find these three examples of different settlement patterns related to land use on the map:
(a) *commercial*: the Apapa port, trading estates and commercial area
(b) *rural-agricultural*: the areas between Badagri Creek and the sea in the south
(c) *a built-up area of urban residential and mixed development*: Lagos Island.

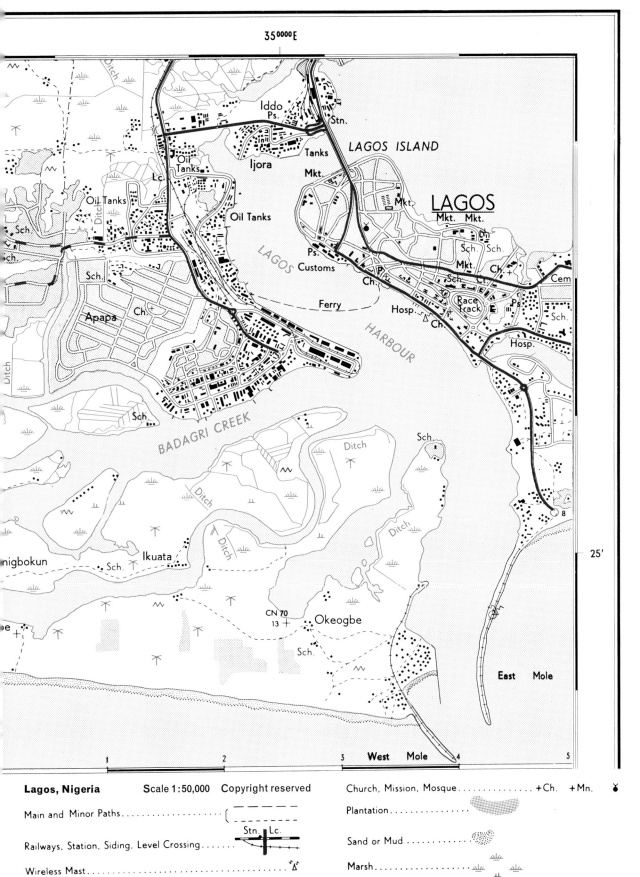

Figure M2 Lagos, Nigeria, 1:50 000

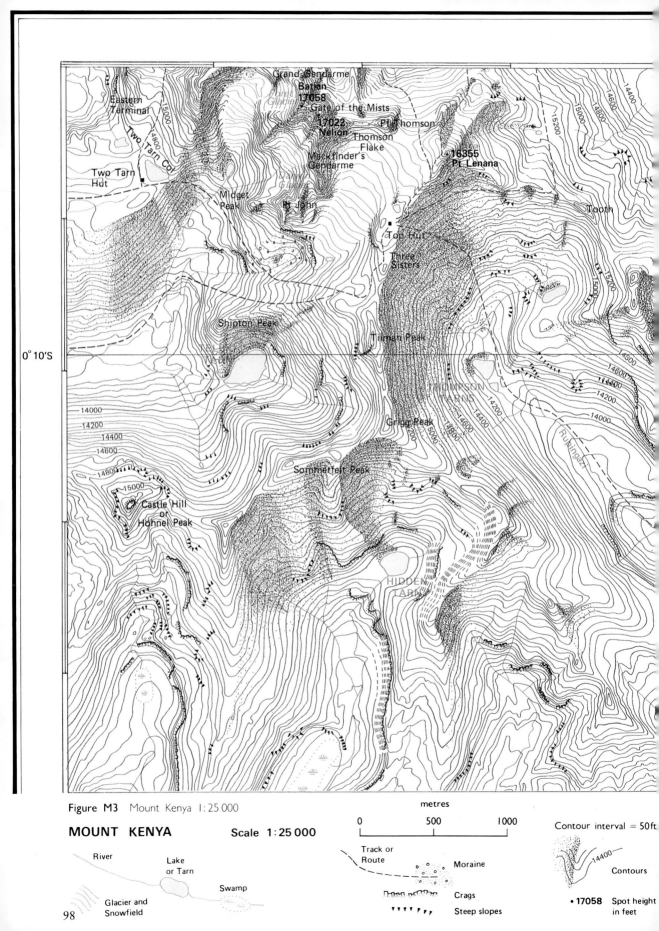

Figure M3 Mount Kenya 1:25 000

The Maghreb 99

Map extract box 2: an ice-worn landscape: Mount Kenya

- The present *glacier* occupies the main valley. It is a massive tongue of ice filling the deep *U-shaped valley*. The valley was gouged out during the glacial period as the great mass of ice moved slowly downhill.
- *Corries* (rounded basins) occupy areas in the tops of the side valleys which were scooped out when they too were filled by ice. They now contain the small *tarn* lakes.
- Further ice deepening of the main valley has left the side valleys at a higher level, as *hanging valleys* with waterfalls.
- These bring eroded material over the cliffs to form *fan cones* at the bottom.
- As the glacier grinds downhill over the uneven rock floor of the valley, the ice flexes, forming deep transverse cracks called *crevasses*.
- The ice melts at the lower valley level forming the glacier's *snout*. Water pours out from beneath, dumping loads of eroded rocky material called *moraine*.

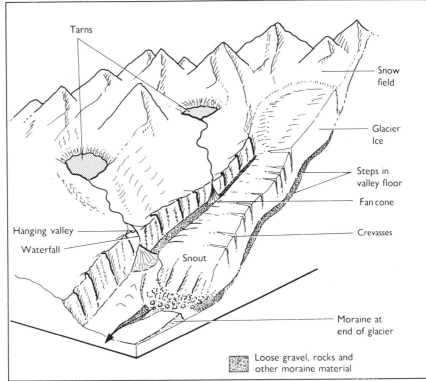

Figure M4 Cutaway section of a glacier, to show glacial features

Study Figure M4, a line drawing and cutaway section of a glaciated landscape and read the explanation.

Now study the survey map extract of Mount Kenya opposite and Figure M3, a line drawing of the same map. Note that the map is at a scale of 1:25,000 and that the spot heights and contours are shown in feet. Figure M5 shows:

- the glaciers and main ridges
- tarn lakes in the high corries (1)
- steep sided straight valleys cut by the ice (2)
- steps in the valley floor (3)
- hanging valleys, with waterfalls and fan cones (4)
- moraine below the snout of the glacier (5)

Identify a good example of features 1 and 4 on the survey map extract. In your notebook, draw a slightly enlarged contour sketch of each, and label it.

Figure M5 A simple line drawing of the Mount Kenya map extract

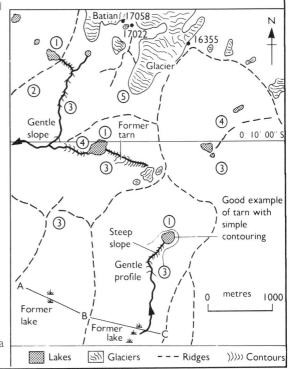

continued from page 95

(markets), palaces, citadels, mosques and barracks. They often have many features in common such as an old city.

The importance of town life in Africa cannot be overestimated. Very often in writing about Africa more emphasis is placed on agricultural development. It is therefore important to emphasise that not all African life is carried on in villages and the countryside, though much is. Most African towns now have electricity, and small workshops and craftsmen; light industrial, service, and transport functions. All are expanding.

Figure 10.2 shows the key economic developments in the Maghreb. Use the map and the trade summaries to answer these questions.
1   Which 2 minerals are important exports?
2   Where are they obtained?
3   How are oil and gas transported to the coast?
4   What are the chief agricultural products?
5   Do any of the countries have well-diversified exports?
6   What is the balance between agricultural and mineral products?

Figure 10.7 A typical scene in the town of Sfax, Tunisia

Algeria

Capital: Algiers 1.7m (1983)

Exports $12,742 million
Imports $9516 million
Visible trade balance $3226 million surplus (1983)

Export commodities	%	Export partners	%
Petroleum	99	France	32
		USA	15
		Italy	14
		Netherlands	12

- Second largest country in Africa, but four-fifths of it is mainly desert.
- One of the richest nations in Africa, because of the production of oil and natural gas.
- Algeria has pioneered the transport of liquid natural gas (LNG) in special ships with refrigerated high pressure storage tanks.
- Petrochemical industries are based on processing 'downstream' products of oil and natural gas. This has protected the country from the worst effects of the reduced world market price for oil.
- The emphasis of development is now on the agricultural sector, as a great deal of food has to be imported. There is usually a healthy trade surplus for this purpose.

Tunisia

Capital: Tunis 600,000 (1984)

Exports $1777 million
Imports $2893 million
Visible trade balance $1116 million deficit (1984)

Export commodities	%	Export partners	%
Petroleum	45	France	21
Clothing	12	USA	19
Fertilisers	8	Italy	18
Phosphates and phosphoric acid	8	West Germany	9
Olive oil	6	UK	4
Fruit	2	Netherlands	3

Morocco

Capital: Rabat 436,000 (1981)

Exports $2161 million
Imports $3569 million
Visible trade balance $1408 million deficit (1984)

Export commodities	%	Export partners	%
Phosphates	24	France	24
Phosphoric acid	19	West Germany	8
Citrus fruit	6	Spain	7
Clothing	5	Italy	7
Preserved fish	3	India	6

- Includes coastlands, High Atlas mountains and Sahara desert.
- The world's leading exporter and 3rd producer of phosphate rock, used to make fertiliser.
- It has controlled Western Sahara, where there are further deposits of phosphate, for 12 years.
- Important sea fisheries (tinned sardines).

Chapter 11 The arid lands of northern Africa

> ## Key words
> Oasis, palmery, aquifer, fossil water and soils, social change

Hot and cold desert occupies 36 per cent of the land surface of the world. This means that a third of the earth is too dry or too barren for easy human living.

The Sahara desert forms by far the largest single tract of arid land in the world. It stretches for over 5,000 km from the Atlantic Ocean to the Red Sea and beyond into Asia; and for nearly 2,500 km north to south, an area of about 10 million sq km. Thus it occupies nearly 40 per cent of the land surface of Africa. If the marginally dry lands are added, the total is nearly 60 per cent, and Africa can be regarded as the second driest continent after Australia. Indeed, lack of water may be one of the most important factors hindering development over about half the continent. There should be serious study of the ways in which the 10 million sq km of arid country in Africa can be used (Figure 11.1)

The arid or semi-arid parts of Africa are found on either side of the northern and southern tropics. In the south, the Kalahari and Namib deserts are relatively small, but the Namib is extremely dry. In the north the Sahara desert is a realm of varied natural landscapes that only rarely owe their character to the work of people. Only about one-eighth of the area is sand-desert; the rest is rock or pebble-desert, or mountainous country. Oases cover about 260,000 sq km or 2·5 per cent of the desert area. The desert area includes the southern parts of the 3 Maghreb countries, Mauritania, and most of Libya and Egypt.

Camels are considered typical of dry lands but were not much in use in Africa until about 300 AD. Before that horses were used. Trade and conquest were probably accelerated as a result of the introduction of camels. The great kingdoms and empires of the Sudan (Ghana, Mali, Hausa, Bornu, Songhai and Darfur) and the Saharan trade routes flourished between the fifth and eighteenth centuries AD. At that time contact between the Mediterranean and southern Europe

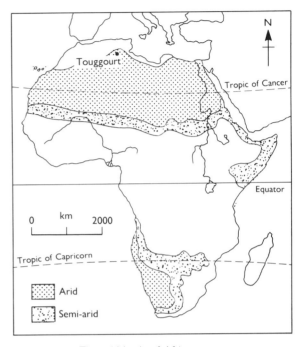

Figure 11.1 The arid lands of Africa

and the Guinea Coast was easier across the Sahara than by sea. From the eighteenth century onwards there were direct sea links between northern Europe and the west coast of Africa.

The pre-historic Sahara

Was the Sahara always desert? There are two kinds of evidence to suggest that there have been changes in climate. One is the size of some of the water-worn valley systems, now no longer occupied by rivers, though occasionally – every 10 or 20 years or so – filled with a flash flood from a storm. The second type of evidence includes abundant indication of life several thousands of years ago.

There are fossil tree-trunks of oak and cedar forests akin to those of Europe, and of savanna trees and rock drawings and paintings of both hunting and herding scenes and of big game animals. Thus physical and archaeological evidence supports theories of climatic change.

People also ask if the desert is increasing in size, and the answer is, almost certainly.

The present climate

The Sahara desert is one of the hottest and driest areas in the world. For much of the year the Saharan high-pressure system repels rain-bearing winds.

Touggourt is situated near the northern limits of the Sahara and the months with occasional rain are the 'winter' months, when 'European' conditions extend southwards over the Mediterranean and parts of north Africa.

Use the figures for Touggourt (see page 24) to obtain the following information:
1 The total rainfall for the year.
2 The months that have no rain. Is this the hot season or the cool season?
3 The months with the highest and the lowest temperatures, hence the average annual range of temperature.
4 How near to freezing are the lowest average temperatures?

Such low rainfall figures are worth almost nothing when daily temperatures are very high. It is all the more remarkable that there should be such concentrations of people within the limits of the desert, (see Figure 11.2) mainly where there are exceptional water resources. The highest population densities in the whole of Africa are found in the oasis of the Lower Nile valley in Egypt (see page 107).

Newer 'oases' include the areas where minerals, especially mineral oil and natural gas, have been found and are being worked.

Fossil water: Savornin's Sea

The key question is 'Why are there such large oases here?' They owe their origin to remarkable supplies of underground water, tapped by artesian wells.

The water gathers in a very coarse sandstone layer 90–2,000 m thick, sealed in by clay or marl layers above and below. This is a porous *water-bearing* rock called an *aquifer*. The water is under pressure, and gushes to the surface naturally when tapped by a bore-hole. The aquifer stretches at least 1,000 km south of the Atlas Mountains and eastwards to the Fezzan in Libya. This vast underground 'reservoir' is sometimes called Savornin's Sea after the French geologist who did so much work on mapping it. The Nubian sandstone, an *aquifer* in Egypt, is possibly the largest underground reservoir in the world.

Clearly the amounts of rain falling over the Sahara today cannot supply these great reserves of water. Geologists are convinced that the supplies have been building up over a very long period, possibly as much as 7,000 years. If this is so, then it is fossil water just as the shells of animals and plants of former ages are fossils. It is a resource that can be exhausted, just as coal or copper or oil deposits can be.

Figure 11.2 The Ziz oases in southern Morocco. The villages occupy the higher ground bordering the palmeries and cultivated land in the valley

The oasis towns at Touggourt

The five main settlements at Touggourt are better called towns than villages. Their total population was 90,000 even in 1931. The houses are similar to Arab-style houses elsewhere; for example, to those of the old town of Kano, in Northern Nigeria. Like homes in most Muslim and desert towns they are packed together for protection and shade and are built round a central courtyard for privacy. Town houses have several storeys and there are electrically lit homes with cool patterned tiles on the floor and often on the walls too.

Touggourt occupies a low bluff a little above the broad valley of the Wadi Rhir. Use the map (Figure 11.3) and the key to answer the questions.
1 Use spot heights to find the valley, that is, the lowest land.

The arid lands of northern Africa 103

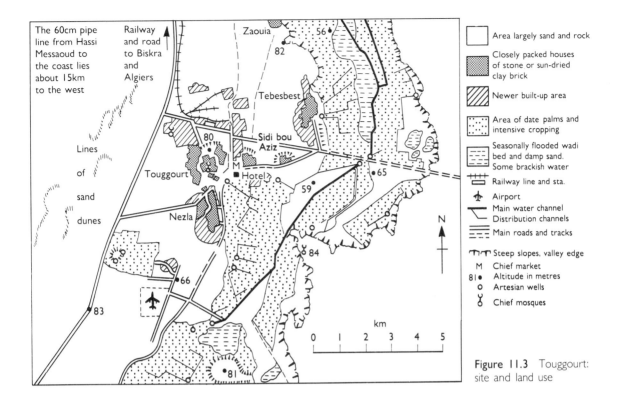

Figure 11.3 Touggourt: site and land use

2 What physical features mark its east and west limits?
3 How does its shape remind you of a river?
4 There is no river in it now. What does the key say about seasonal water?
5 Measure the distance from one side of the valley to the other; what does this suggest about the size of the river and the climate when it was formed?
6 Is the palm-growing area in the valley or on the higher ground? Give possible reasons for this.
7 Are the towns placed in the palmery or not? Suggest reasons why the towns are usually placed on the higher ground.
8 What different kinds of communication link are shown? What others are there likely to be?

The market and its products

On a full market day the market place of Sidi bou Aziz is packed with people. The goods for sale are bought by townsfolk and nomads, people from outlying villages and from the fringes of the oasis where some nomad people are setting up permanent homes.

There were date palm saplings and long sprays of waxen date palm flowers for sale in the market in April.

• The date palm saplings are shoots taken from wild palms that have not been irrigated. They are hardier and can be set in circular pits and left to survive.
• They will not bear fruit for between 6 and 7 years.
• The date palm flowers on sale are male flowers. Male trees do not bear dates and most of them have been cut for building timber, fuel, and for lining wells. Hence the proportion of male to female palms is now only about one to 30 or even 50 productive female trees. This means that there are not enough male trees producing flowers for natural pollination by insects and the wind.
• A well-pollinated head of flowers produces up to ten pounds of dates; an insufficiently pollinated head only one pound. One method of pollination is to tie a small sprig of male flowers in the centre of the branch of female flowers before it opens so that pollination occurs as it comes to full bloom.
• One of the new developments in commercial palmeries is pollination by a spray blower.

Look at the practical workbox on page 104.

Practical work box 7:
how to answer a resource based examination question on oasis farming

If suitable source materials are given, you can answer this type of question even when you have not learned about this particular example. This question asks about the *basic facts* of oasis farming, for example, that irrigation water must be *raised* to flow by gravity, and crops must be *drained* or they will become waterlogged.

A palmery near Touggourt
Try to answer the following questions based on the line drawing, Figure 11.4 and the photograph, Figure 11.5.

	Marks
1 Why is it possible to farm this land so intensively? What means are used?	5
2 Why is the main pipe raised up on an earth wall?	1
3 Why is it necessary to have the ditch?	1
4 Give two or three reasons why the crops marked are grown in the palmery.	2
5 Why is barley grown in the open near the edges of the palmery?	1

Answers might be:
1 Multi-cropping: several crops a year; multi-storey cultivation; irrigation; irrigated seed beds; transplanting; economical use of ground area
2 So that water can flow naturally to the chosen area
3 To drain the ground
4 Annual vegetables are grown for local use and early markets; perennial tree fruits and dates share water and give shade.
5 It does not need shade or much water: a winter crop.

You can test yourself by practising on other photographs in the book, for example, Figure M6, Nkana mine on page 130. Could you answer a question on mining, having studied copper or coal or gold? It would include questions on simple geology, mining methods, processing, transport and infrastructure.

Figure 11.4 A line drawing to show multiple cropping in the palmery

The arid lands of northern Africa 105

Figure 11.5 A part of the palmery at Touggourt in early April

The mineral wealth of the desert

The desert also holds great mineral resources. Rock folds and structures that can conserve water may also conserve mineral oil and natural gas if the conditions for their formation are right. The first strikes were at Hassi Messaoud in Algeria in 1956 at a depth of over 3,000 m, that is, 3 km below the surface. Now other Saharan fields are producing on the borders of Tripolitania and further east in Libya and Egypt. Both gas and oil are carried by pipelines from Hassi R'Mel, nearly 500 km due south of Algiers to the sea and across the Mediterranean to Europe. Special methane tankers carry gas to northern Europe in liquid form where it is restored to normal temperature and fed into the gas pipeline distribution system.

There are valuable deposits of other minerals

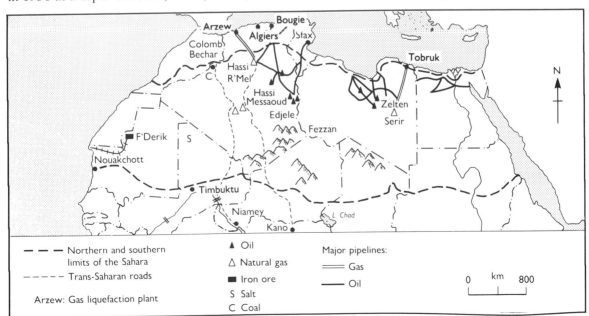

Figure 11.6 The Sahara: minerals and transport links

in the desert. The problem is accessibility. The vast iron ore mountain at F'Derik, far to the west in Mauritania, has been linked to the coast by a railway 650 km long. The ore has been the main export of the country, but supplies have been disrupted by the activities of Polisario, the Western Sahara liberation movement. Look at the trade summaries for Libya and Mauritania to see how important minerals from the desert are to the economies of these countries. Check the map, Figure 11.6, to see how these minerals are transported.

Libya

Exports $10,056 million	*Capital:* Tripoli, 991,000 (1984)
Imports $8,079 million	Visible trade balance $1,977 million surplus (1984)

Export commodities	%	Export partners	%
Petroleum	100	UK	27
		Italy	24
		West Germany	10
		Spain	7
		Turkey	5

- Largely desert and semi-desert. Population concentrated in the two main cities the coast.
- Oil production has transformed the economy. The GNP is the fifth largest in Africa, but is shared among a small population, giving Libya the highest per capita income in Africa.
- Oil revenues have been used to improve infrastructure and social facilities, and to use fossil water supplies to expand local agriculture.

Mauritania

Exports $315 million	*Capital:* Nouakchott, 350,000 (1984)
Imports $378 million	Visible trade balance $63 million deficit (1983)

Export commodities	%	Export partners	%
Fish	50	France	27
Iron ore	50	Spain	11
		Other EEC	37

- A thinly peopled semi-desert country where the main activity has been nomadic cattle herding.
- Dust storms, clearly seen on satellite pictures, have been increasing since the 1960s. Every year 100 million tonnes of topsoil is blown away.
- Iron ore mining in the northern interior is the mainstay of the economy.

The breakdown of traditional ways of life in Algeria and Mauritania

Algeria's present prosperity has been achieved in spite of the total disruption of the country, and especially the rural way of life, during the struggle for independence. The development of resources in the central Sahara has brought immense changes both in the traditional oases and in the new mining oases.

The transition is sometimes slow. People who formerly held grazing rights are being dispossessed. Some of the nomads have settled. Men from the oases or from nearby pastures earn a cash income by seasonal work picking dates, or grapes in the vineyards of the coastal valleys of northern Africa, or as carriers, or labouring in one way or another on the new enterprises.

In Western Sahara and Mauritania the breakdown of traditional ways of life has been even more devastating. Almost permanent drought since the late 1960s and early 1970s and years of over-grazing have resulted in the desert moving south and west at a rate of 5 km a year. Four-fifths of the country's grazing lands have disappeared.

> 'A country blowing away across the Atlantic'

- For the fiercely independent nomadic herdsmen who have lived here for centuries this has meant a hard choice. 'If we stay in the sand we die. If we go to the cities our society will die.'
- Refugee camps have grown up near large towns and along main highways and rivers.
- Nouakchott was built as the capital of a newly independent state in 1960 for a few thousand people. It is now reputedly the largest refugee camp in Africa. By 1982 it had a population of 250,000; by 1986, 600,000.
- For most of these people there is little hope of returning to their traditional way of life.

North-eastern Africa

Chapter 12 Egypt and the Nile Valley

Key words
Water control, consolidation, food-population gap, satellite cities, subsidies

Agriculture in the Nile Valley

Most of north-eastern Africa would be part of the Sahara desert but for the River Nile. But the Nile is not just a water resource: it is also a great power resource and means of communication.

Transport was hindered by the 6 cataracts between Khartoum in the Sudan and Aswan in Egypt. The cataracts are formed by ribs of hard granite. All this has changed. The First Cataract now provides the foundation for one of the world's great dams, the Aswan High Dam. The water of Lake Nasser covers the Second Cataract and reaches the Third Cataract, providing a long stretch of navigable waterway.

The water of the Nile is a vital asset to the economy of the 4 major countries that are directly concerned with it.

The greatest oasis
The area drained by the Nile is vast and, during its 6,000 km course, the river crosses regions so different that in August and September when cotton is being planted in the Sudan it is being picked in Egypt over 1,500 km to the north. It is one of the most important rivers in the world because for half its length it flows through desert, turning parts of it into an immense oasis.

The green and settled area forms a narrow belt along the main valley of the River Nile, sometimes 3–6 km, sometimes as much as 15 km wide.

Although the river swings from side to side it is hemmed in by steep cliffs of Nubian sandstone for hundreds of kilometres of its length (see Figure 12.1). These have restricted the area of flood and silting and in the past made it easier to irrigate the valley land.

Within the narrow valley there is bustling life in villages and towns. Large cotton ginneries,

Figure 12.1 The temple at Abu Simbel before the flooding of Lake Nasser. The grid of small fields were farmed, and temporary homes used until the flood came every year in September. The temple has now been raised and resited at the top of the cliff

power plants and other factories are found in towns linked by rail to Shellal (Aswan) in the south and Cairo and Alexandria in the north. This 800 km long narrow strip along the river carries population densities of 1,000 people to the sq km.

The Nile flood
The annual flood gave life to the land for thousands of years.

Although Egypt is a country of a million square kilometres, the arable area which supports 96 per cent of the people is under 3 million ha, about 3 per cent of the total area. This is due to irrigation and the silt brought down by the annual flood.

• The 2 main source areas for water are the Blue Nile rising in Lake Tana in Ethiopia, and its tributaries; and the White Nile and its

tributaries rising in the mountains, plateaus and lakes of East Africa.
- They supply water at different times of the year.
- The Blue Nile provides two-thirds. When the Ethiopian rains of March and April reach Khartoum in August and September the Blue Nile is so strong that it 'holds back' the White Nile.
- The White Nile has a more even flow but supplies less than a third of the water. It has in the past supplied water to Egypt in spring and early summer when water is desperately needed.

There are also natural aquifers beneath the sands of the western desert, similar to those described in Chapter 11, page 102. These are fed partly by underground seepage from the River Nile. The water-bearing layers are very deep, probably from 1,700 to 3,500 m and have the great advantage of natural annual renewal and low salinity (saltiness).

The Egyptian government hopes to irrigate a million hectares of *new* farmland by the end of the century. This would need 12 billion cubic m of water more than the 55 billion already provided by the River Nile. They hope to achieve this by:
- Tapping a further 2 billion cubic m of groundwater.
- Reducing water loss by evaporation and percolation in the Sudd swamps in the Sudan when the Jonglei canal by-passes them. Construction was started in 1978 but has been halted by anti-government activity in the Sudan (see Figure 12.2).
- Recycling more of the 16 billion cubic m of drainage water that is lost to the Mediterranean and delta lakes each year.
- More economical use of water in existing irrigated areas and other water saving projects higher up the Nile.

Water saving schemes such as these are particularly important when drought reduces Egypt's water stocks in Lake Nasser to dangerously low levels.

Traditional methods of irrigation

There is an immense contrast between the small scale and the large-scale use of water for irrigation. Figure 12.3 shows different irrigation methods and Figure 12.4 shows different scales.

Figure 12.5 shows land and villages bordering the River Nile. The example is taken from the Sudan because it shows what settlement and farming were like before the introduction of

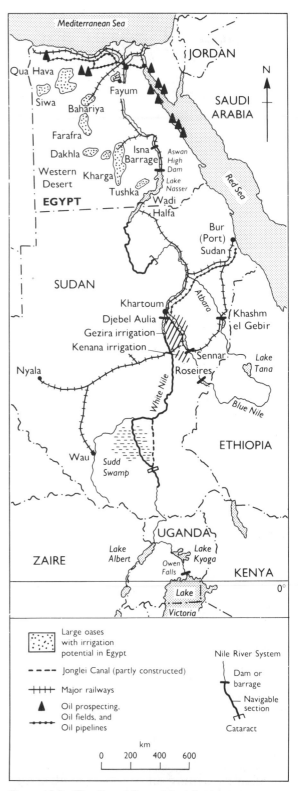

Figure 12.2 The River Nile and its tributaries

Egypt and the Nile Valley 109

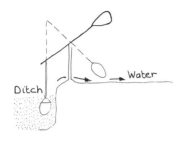

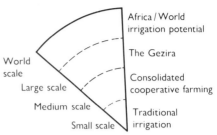

Figure 12.4 Irrigation at different scales

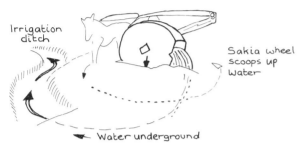

Figure 12.3 Traditional methods of lifting water for irrigation: a shaduf and a sakia wheel

large-scale perennial 'all-the-year-round' irrigation from reservoirs.

Detective work on Figure 12.5. Find answers to the following:

1 First read the caption and then study the key. Why does it say that the heights of the land are 'heights at that time?'
2 The numbers in circles on the map show land use and methods of irrigation. Check the numbers with the symbols shown in the key to make sure that you understand them. For example, number 1 is land irrigated by sakia

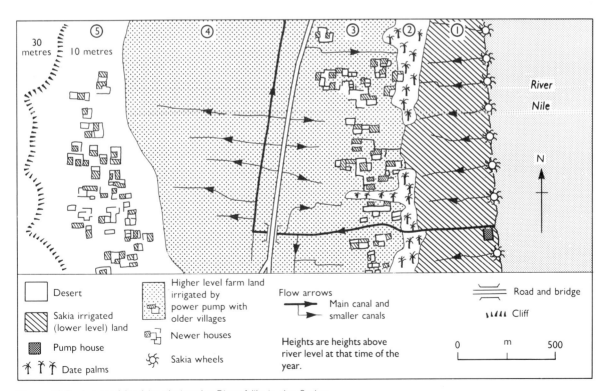

Figure 12.5 Irrigated land bordering the River Nile in the Sudan

wheels. Then make a list giving the name of each of the five contrasting zones.
3   What tells you that the main canal is higher than the cropped land?
4   The first stage of settlement covered areas 1, 2 and 3. There was no main canal at that time and the homes were on unirrigated land. Give 2 reasons why they were not placed nearer to the river.
5   In what ways has the settlement story repeated itself as the population increased and more homes were needed?

The shaduf and other traditional devices are still used to lift water. The tambour lifts water only about one metre but the water wheel or sakia can raise water about six metres. Modern sakias are no longer creaking frames with clay pots tied round the edge of the wheel. They are mass-produced iron wheels which can be slotted on to fixed cement posts and moved as required.

The term 'basin irrigation' was given because once a year at high water the river was allowed to flood into carefully prepared basins.

Although water may be allowed to flood small sections, irrigation more often means ridging a field so that water flows down the furrows between the rows of crops planted on the ridges, so that it sinks into the ground to supply the roots (see Figure 12.6). It involves making, breaking and re-making small earth barriers (termed *sadd*) so that water flows down each section in turn; it involves a whole series of regulators and flow points, and sluices of different sizes in order to exercise proper control.
• Water travels along the main irrigation channels which keep to the high ground so that branch canals and distributaries can lead water by gravity flow to the maze of smaller channels graded to the natural slope.
• Water has to be shared. There are irrigation 'turns' of 4 or 5 days followed by the same length of dry time; or sometimes a double length of dry time. Much depends on the crop.
• In Egypt wheat needs three or four waterings during its life, cotton twice as much as that, and rice most of all. Regulating the water is extremely difficult when many farmers independently decide what crops they will grow on many scattered plots.

Almost always irrigation involves cooperation. If one man is lifting water by *shaduf* and pouring it into a channel, another man 30 or so metres away is closing one channel with an earth wall

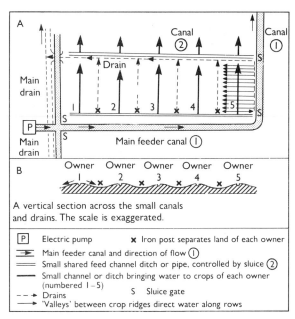

Figure 12.6 Watering plan for irrigated fields in a cooperatively farmed area

so that the water will take a different path. If two men are operating a *tambour* (or Archimedean screw), see Figure 12.7, someone else is ensuring that the water flows to different parts of the field in turn. It is the same if a *sakia*, turned by a blindfolded bullock, lifts the water. It is equally necessary to cooperate in sharing water on a large scale.

Increasing agricultural output:

Egypt's population has been increasing by about a million people a year (2 per cent). More food is needed not only to feed these extra people but also because people's expectations about the quantity and quality of their diet have risen.

It is very difficult for Egypt to increase agricultural output.
• The green Nile tract is less than 3 per cent of the area of Egypt, but 96 per cent of the people live there. Population densities are around 1,000 per sq km or higher.
• Yields of cotton, maize, wheat and rice are already among some of the highest in the world. Rice output has already doubled. It is difficult to go on pushing up yields.

Egypt and the Nile Valley 111

Figure 12.7 Archimedean screw. The interior is like a large corkscrew. The spiral coil 'winds' water from the lower end (in the main canal) up to the small ditch taking water to the fields. Note the support wire. It is worked by two men sitting on the bank, turning handles at the top

- Almost all the tillable land along the 1,200 km length of the Nile in Egypt is already converted to perennial irrigation and multi-cropping.
- The area of cultivated land is still under 3 million ha. Although more land has been irrigated, it is balanced by loss of land to towns, industries, new roads or airports.

Efforts to increase production are being made in several ways.
- Providing more water for irrigation.
- Reclaiming land, for instance by levelling in the Fayum depression, and by infilling lagoons near the coast with river silt.
- More intensive use of the land by multi-cropping (that is, 2 or more crops on each piece).
- Introducing better varieties of seed to increase yields.
- Making more efficient farms by regrouping land holdings. This is called land consolidation.

Balancing people and production in present-day Egypt

There have been repeated statements over the last 30 years about the difficulty of keeping agricultural production in Egypt in step with rapid population increase.

In 1962 President Nasser said: 'It is the most dangerous obstacle that faces the Egyptian people in their drive towards raising the standard of living in their country.' Twenty years later it was said that 'Increased population growth has changed a food surplus into a deficit.' A newspaper headline reported: 'Baby boom adds to the economic plagues of Egypt.'

In Egypt there are one million more mouths to feed every year. At this rate the present population of 46 million will be over 70 million by the year 2000 and 100 million by 2015.

In some years, one-third of Egypt's food has to come from overseas, including wheat, cooking oils, and beef for a Muslim country that does not eat pork.

The total 'food gap' costs Egypt $4 billion a year. Egypt has doubled its rice crop, but there is still a rice gap, and also a sugar gap as well as more demand for the foods mentioned above.

The Egyptian government has tried to cushion this hardship for the poorest people by subsidies for many basic commodities. But subsidies benefit everyone, the rich as well as the poor.
- The government is afraid to reduce the subsidies for fear of food riots (as happened in Zambia).
- All this is only possible with massive overseas aid and loans (see Figure 12.9).

The result is that Egypt has a $30,000,000 foreign debt and is desperately trying to:
- limit and keep pace with the 'baby boom'
- pay off loans and interest on loans
- increase export commodities to earn foreign money to pay for food imports
- spend vast sums on projects that will increase production.

Look at the trade summary for Egypt and note that it has a large visible trade gap. Foreign income to offset the deficit comes from:
- remittances: up to 3 million overseas workers

sent back $3.8 billion in 1983
- Suez Canal dues from shipping earn $1 billion a year
- tourism which earned $386 million in 1982 but has dropped by about one-third because of fears about terrorism

The following sections explain some of the ways in which Egypt is investing in capital projects to help increase production. They include Aswan irrigation and power projects, Aswan fisheries, Lake Nasser and oasis irrigation projects.

Egypt	*Capital:* Cairo, 12 m (1984)
Exports $3,693 million	Visible trade balance
Imports $7,515 million	$3,822 million deficit (1983)

Export commodities	%	Export partners	%
Petroleum	57	Italy	18
Cotton	23	France	9
Fruit	3	USSR	7
		USA	7

The Aswan High Dam

Egypt's most ambitious project has been the planning and building of the Aswan High Dam.

The new Aswan High Dam, 3 km wide and 111 m high, has created a narrow lake, Lake Nasser, 500 km long (see Figure 12.2). It has

Figure 12.9 Aid for Egypt: who benefits?

about 25 times the capacity of the old Aswan Dam, and irrigates 500,000 more hectares.

In the autumn of 1965, perhaps for the first time in a million years, the Nile was not allowed to flood. In the past, 80 per cent of the flood water came between August and October, leaving only 20 per cent for the remaining 9 months. The present High Dam aims at regulating water for year-round (perennial) irrigation. In addition it produces electrical power (45 per cent of Egypt's requirement) both for the industrial sector, and for rural electrification. Fisheries have developed in Lake Nasser and it is proving a tourist attraction.

At the beginning there were problems.

Figure 12.8 A village in the Nile delta in September. The cotton crop is almost ready for picking

- 500,000 people had to be resettled.
- Lake Nasser would flood unique archaeological sites.
- There were fears that Lake Nasser would fill up with silt and become useless, and that the loss of silt to the land would lower farm yields.

1 *Moving people*

People from the Wadi Halfa area in the northern Sudan have moved to a new project 800 km to the south-east at Khashm el Girba on the river Atbara.
- Most of the Egyptians have moved into about 75 government townships, the largest of which is Kom Ombo, 65 km north of Aswan. Nasser City is the new Nubian capital.

2 *Archaeological sites*

There was an international effort to save the archaeological treasures that would be covered by the rising waters of Lake Nasser (see Figure 12.1).

3 *Silting*

Engineers estimate that it will take about 500 years to fill with silt.

There are still some problems.
- Because of the year-round irrigation, the water table has risen in some places, making the ground waterlogged.
- Because there is no annual flood, fertiliser, chemicals, and effluent from towns and industries do not get washed out, so salinity is increasing.
- In the same way canals are not washed through and do not dry out every year, so there are more weeds and more snails to host the illness called bilharzia (see Part 3).

Developments at Aswan

As in many other parts of Africa where huge projects are under way or completed, the old Africa is almost unrecognisable.

The developed Aswan is both an industrial and a tourist city. Abundant electrical power makes possible the establishment of chemical and more nitrate fertiliser factories, by nitrogen-fixation from the atmosphere. The latter are needed to replace the silt, for the Nile will no longer flood at will over the land of Egypt. Steel mills, electronics plant and traditional industries may in time justify the suggestion that Aswan will become the 'Pittsburgh' of Egypt.

Lake Nasser fisheries

The Lake Nasser fisheries are a success story but the catches are still far below the totals that Egypt could use. Fish is an important source of food and costs half the price of meat or poultry in Egypt.

The fish species are river fish living in surface water. Two main methods are used to catch them.

1 Simple gill nets about 30 m long, with small holes, are used. Nets are pulled in at sunrise after overnight fishing, and the catch placed on ice.

2 Larger Nile perch (tilapia) are caught with nets 150 m to 200 m long and about 1.5 m deep. The boat circles so that the net forms an enclosure and the fish are trapped.

These methods are helped by the fact that the lake *stratifies* (forms layers of water with different light, heat, salinity, density and food supplies) in summer. The fish stay in the upper level. Peak fish landings are March and April, and October to January.

> 'Fish could feed more people at lower cost: it would help fill the food gap and reduce imports'

Carrier boats call daily and collect catches; mother ships store fish on ice and market it every 2–3 days. Floating refrigerator barges for storage and icemaking and better harbours are needed.

Oases in the western desert

Egypt is 'thinking big' about ways of using some of the oases in the western desert. There are at least five areas that might be developed to bring more land into cultivation and increase food output. The problems are those of distance and isolation from the main valley; and of providing irrigation water. They have an advantage in that they are depressions, often below sea level (see Figure 12.2).

One area, El Fayoum, is the most highly developed, partly because it is only 90 km south-west of Cairo. It has a population of nearly 3 million. El Fayoum is a depression about 45 m below sea level and its natural water supply is supplemented by irrigation water brought by canal branching from the Nile near Assiut. The oasis is large, nearly 700 sq km. Only a small lake now remains of the larger one that at one time filled the depression. The soils are fertile and produce 2 crops a year of cotton, cereals (maize, wheat, barley, rice), vegetables (tomatoes, onions, potatoes), fruit (citrus, dates, figs). This rich variety is typical of desert oases provided that the land can be irrigated.

Other oases are larger still, but await development. A string of oases lies about 200 km to the west of the River Nile, and parallel to it. The largest depression of all, the Qattara, is further north. For decades there has been talk of cutting a canal to it from the Mediterranean Sea.

Town life and industrial development

We often think of Egypt as a rural and agricultural country. Yet nearly half the people live in cities, towns or large urban villages, the largest of which is Cairo.

Cairo, the largest city in Africa

Cairo is the focal point of activity in Egypt. It is the largest city in Africa, the largest city in the Middle East and the largest Islamic city in the world. Its population is probably about 12 million though the official estimate of the population is 8.5 million (1980s). It is likely to reach 20 million by the year 2000 if present trends continue. The Cairo–Alexandria–Delta–Suez zone is now being described as a future *megalopolis*. It already has 60 per cent of Egypt's urban population even though it is still one of the most intensively farmed areas in the world.

Use Figure 12.10 to check reasons why the city may have developed at this site.
1 The islands help to make bridge-building easier.
2 Cairo is on the right bank. How is the relief here different from that on the left bank?
3 Giza is on the left bank. Note the width of the zone of farmland.
4 The built-up area stretches out north-eastwards towards the new airport and Suez.
5 The valley is limited on both sides by cliffs of Nubian sandstone. It narrows near the site of the citadel and Old Cairo.

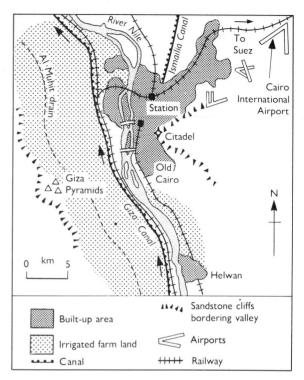

Figure 12.10 The site of Cairo

As in other great cities, expansion is causing tremendous difficulties. There are the usual problems of traffic congestion, water and power supplies, drainage, and telecommunications, as well as the shortage of homes. Some people are living in the 'City of the Dead', one of the large cemeteries. There are infrastructure improvements, including new sewer systems, roads and an underground railway. However, the city has become so large that it is virtually impossible to make it work properly, especially when money is short.

To reduce the pressure on Cairo, the government is expanding the 3 cities alongside the Suez Canal: Suez, Ismailia, and Port Said. Each of these is to grow by 250,000 or 500,000 people. There are also plans for 5 satellite cities around Cairo (see Figure 12.11). These are to be situated about 30 to 90 km from Cairo, and could accommodate up to 3 million people. New city building on this scale will be an enormous undertaking.

There are two kinds of new town: 'dormitory' towns, for example a large workers' town in the desert just outside Helwan; and others which are intended to become full scale independent cities, developing their own light and heavy industries.

An example of the second kind is Sadat City, half-way between Cairo and Alexandria. A similar development occupies a site 50 km along the road to Ismailia.

Industrial development of all kinds is being promoted in Egypt to provide jobs, and to reduce imports. Egypt is the second most populous country in Africa after Nigeria, and has an important home market. It also has a key position for eastern Mediterranean as well as Middle East markets.

The expansion of industrial production and employment has been faster than that of agriculture in recent years. The Delta–Suez area is the natural choice for industrial development. It has about two-thirds of the agricultural production of the country, and two-thirds of the population, including Cairo (12 million) and Alexandria (5 million), the two largest cities in Africa. There are new industrial projects in progress around Cairo and near the Suez Canal. The canal itself was opened in 1869, and reopened in 1973 following blockages resulting from the 1967 Arab–Israeli war. Major improvements have been carried out to accommodate ships of up to 150,000 tonnes, and further enlargements for ships up to 260,000 tonnes are in progress. Figure 12.11 shows the importance of the canal zone to Egypt and also Egypt's strategic position in Africa.

Egypt's industrial growth has been as high as 15 per cent a year and averages 9 per cent a year, a considerable achievement. There is a good resource base, with deposits of iron ore, coal,

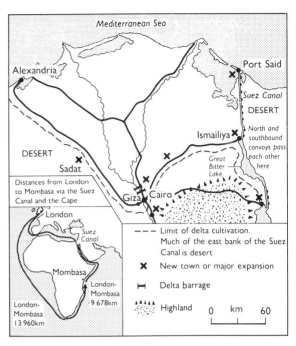

Figure 12.11 The Nile delta (inset – Suez in Africa)

mineral oil and natural gas, phosphates, and some metal ores. There are heavy industries such as the iron and steel of Helwan, chemicals, petrochemicals, cement and textiles and nitrogenous fertiliser at Aswan; and widespread food and clothing industries. The last two are profitable exports.

Chapter 13 Sudan, Ethiopia, and Somalia

Key words
Irrigation, drought, disruption, famine

This vast area of north-eastern Africa consists of the Ethiopian highlands, and the surrounding expanses of desert and semi-arid scrubland that cover most of the Sudan and Somalia (Figure 13.1). These two semi-desert countries have a very low density of population, and limited areas suitable for the cultivation of crops. The uplands of Ethiopia, with mountain peaks rising to 4,500 m and large areas between 1,500 and 3,000 m, have a very different character. Average rainfall of 1,200 mm supports extensive coffee cultivation and a much higher density of population. Yet all 3 countries share problems of political conflict and unreliable climate. The table on page 119 summarises and compares the countries.

Of the 3 countries, the Sudan is the most fortunate with the highest GNP per capita and lowest trade deficit. It depends for most of its income on a huge cotton growing 'island' near the centre of the country. The promontory of land between the Blue and White Nile rivers is called the Gezira. It has given its name to one of the earliest and most remarkable agricultural projects in Africa.

Cotton in the Gezira

Flying south from Khartoum to Wad Medani, for 150 km the flight path crosses a treeless plain covered by a check pattern of crops and fallow stretching into the distance (see Figure 13.2). About 1 million ha are in irrigated cultivation, growing cotton and food crops in what is otherwise virtually a desert. This is the Al-Jazirah (Gezira) scheme. It is one of the most impressive man-made sights in Africa. It is significant because it is organised on large-scale lines but is worked in small units by tenant farmers. The more recently developed area to the south is called Managil.

Figure 13.3 shows the location of the irrigation

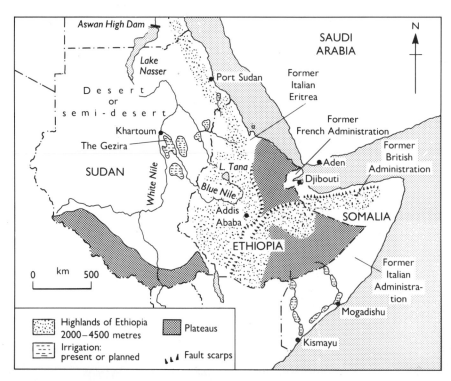

Figure 13.1 The Sudan, Ethiopia and Somalia

Sudan, Ethiopia, and Somalia 117

Figure 13.2 Gezira irrigation. The dark soils are almost level because they were deposited in a former lake. Note the series of irrigation canals and channels, and work out the sequence of water movement from one channel to another (see Figure 13.4)

projects. The following are the important points to note about the Gezira project.
• The irrigation water comes from the Sennar dam, built in 1925 (see Figure 12.2).
• The Gezira has dark grey or black soils formed of fine silt brought down by the rivers from volcanic areas in Ethiopia. These soils can retain irrigation water better than sandy soil.
• Each tenant farmer cultivates his holding on a rotation system, with one-third usually used for cotton. The holdings are arranged in groups so that the water management system can serve several holdings at once (see Figure 13.4).
• The whole scheme is a remarkable partnership combining state ownership and individual enterprise.
• The profits are shared, with a portion going to local government and community development.
• The tenant farmers and their families live in a series of villages. These have a bore-hole well to supply drinking water, a market, a mosque, shops, schools, and a clinic. Firewood plantations are confined to a few locations. More woodland would give shelter to birds which eat the millet and young cotton buds.
• Critics of the Gezira scheme say that it displaced local people and disrupted traditional life, and exploited the workers.

Now study Figure 13.4, Block 10, Wad Medani, to find out more about how one small part of the scheme is operated.
1 Make a list of *all* the names shown on the diagram.
2 Regroup these under the following headings:
• names of settlements
• amenities and social services, including training and education
• cotton growing and processing
• transport

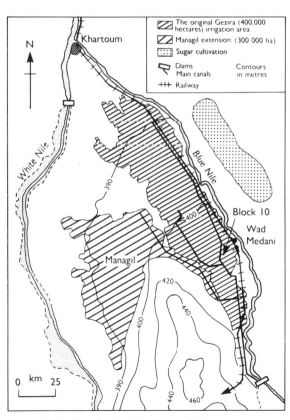

Figure 13.3 The Gezira: irrigation projects

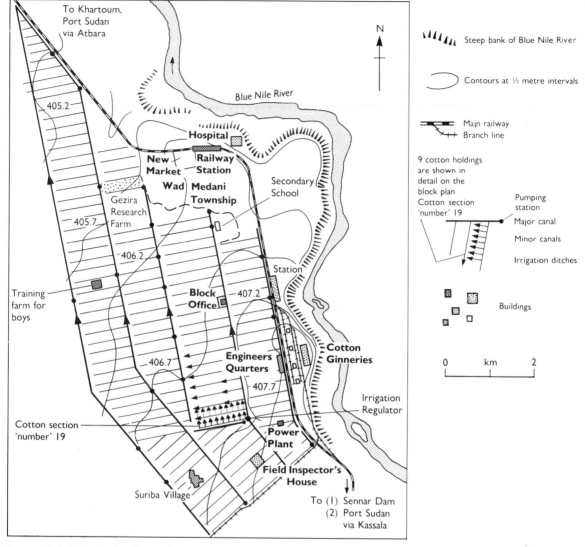

Figure 13.4 The Gezira: Block 10, Wad Medani

3 Study the contours and find out how much the land falls from south to north.
Check the facts about growing cotton in the cotton fact box on page 120.

There are other major agricultural projects in similar areas in Sudan that combine good soils and gradients suitable for irrigation. The most important are:
• Kenana, south of Gezira and Managil, 400,000 ha irrigated from the Roseires dam
• Ar-Rahad, east of the Blue Nile and to the south-east of the Gezira, where more cotton is grown
• Khashm el Girba, where sugar is grown by people resettled from the Sudan part of the Nile valley flooded by Lake Nasser
• Kassala, where the Gash river emerges from the mountains of northern Ethiopia

These areas are served by a railway leading north to join the long-established Khartoum-Bur Sudan (Port Sudan) line.

Common problems in 3 countries

	Sudan		Ethiopia		Somalia	
Area	2,506,000 sq km		1,224,000 sq km		638,000 sq km	
Population	20.6 m (1983)		42.1 million (1984)		4.6 million (1983)	
Density	8 per sq km		35 per sq km		7 per sq km	
Capital	Khartoum		Addis Ababa		Mogadishu	
	(817,000 – 1983)		(1.4 m – 1980)		(350,000 – 1973)	
GNP (1983) Total	$8,420 million		$4,860 million		$1,140 million	
Per head	$400		$140		$250	
Exports Total	$519 m (1984)	%	$417 m (1984)	%	$41 m (1984)	%
	Cotton	24	Coffee	61	Live animals	79
	Sorghum	22	Hides	10	Bananas	7
	Live animals	11	Pulses	4	Myrrh*	6
	Groundnuts	10	Oilseed	3	Petroleum	3
	Sesame seed	8	Live animals	1	Fish	2
	Gum arabic	8			Hides	1
Imports Total	$546 m (1984)		$849 m (1984)		$466 m (1984)	
Visible trade balance	$27 m deficit		$432 m deficit		$425 m deficit	

*Myrrh is an aromatic wood.

There are 2 hazards which are common to all 3 countries in the table above:
1 Low and unreliable rainfall.
• If there are a few good years, larger areas are cultivated and herds get bigger. Harvests are good and people eat.
• When the rains fail, the harvests are poor and the animals die. There is not enough food.
2 Political conflicts based on religious and ethnic divisions have resulted in wars which cause human suffering, refugees, disruption of agriculture and economic development, and a waste of scarce resources.

Sudan

• Northern and western Sudan is desert and semi-arid land, with scattered grazing and cultivation.
• Irrigated agriculture is confined to the valleys of the White and Blue Nile, the Atbarah and other rivers.
• The northern two-thirds of the country is administered as a Muslim state.
• The southern third is part of Black Africa, and follows Christian or traditional animist religions.
• The south has been in a state of armed rebellion against the Khartoum government for decades.
• This has disrupted the digging of the 240 km Jonglei Canal, which would allow the waters of the White Nile to bypass the marshlands of the Sudd.

Ethiopia

• The high mountains of Ethiopia receive adequate rainfall which supports coffee growing.
• There is a desert and semi-arid zone along the Red Sea coast.
• Drought in the north-east part of the country has caused devastation.
• Ethiopia has been largely a Christian country since Roman times.
• The population of the semi-arid coastal province of Eritrea is largely Muslim, and does not consider itself to be part of Ethiopia.
• Eritrea was added to Ethiopia by a resolution of the United Nations in 1951 to give the country a direct outlet to the sea.
• Fighting has occurred periodically since then, and many Eritrean refugees have fled into Sudan.

Somalia

• Most of Somalia has low rainfall, and most people are semi-nomadic pastoralists.
• About one-third of the Somali people live outside Somalia in the eastern part of Ethiopia (the Ogaden) and the north-east frontier district of Kenya.
• The desire for a unified Somali homeland has led to some guerilla warfare in Kenya, and a full-scale war between Somalia and Ethiopia (1977–82) for control of the Ogaden.
• About 1 million Somali refugees from the Ogaden have moved into Somalia.

> **Facts: Cotton**
>
> *Climate and cultivation*
> Prefers temperature of 16–24°C
> Altitude up to 1,400 m
> 750 mm of rain in growing season or irrigation, reliable dry season for picking.
> Fertile well-drained soil
>
> *Processing*
> Lint separated from seed at local ginneries
> Lint baled and sent to textile factories
> Seed sent to local oil mills for extraction
>
> *African production*
> 20 African countries export cotton totalling 18% of world exports. Egypt is 3rd and Sudan 6th world exporter.

Drought + war = famine

The table on page 119 shows that wars have shattered these 3 countries for many years. War is perhaps worse than drought, because it is man made. It creates thousands of refugees with no food, no animals, no homes, and no hope.
It disrupts:
• local food production
• movements to and from grazing land
• development and maintenance of capital projects
• transport of all kinds
• distribution of relief supplies to drought stricken areas

This region, like the Sahel (pages 78–81), suffered severe droughts in the early 1970s and mid 1980s. Harvests failed and widespread famine caused thousands of deaths. This must have happened many times before, but television pictures showed for the first time, to millions of people around the world, what drought could mean to ordinary people in Africa. There was a huge international response, but difficulties and delays in organising relief supplies, and bringing assistance to the stricken areas meant that help

Figure 13.5 Controlling water distribution in the Gezira. The photograph was taken in September when there was plenty of water coming from the Ethiopian highlands. In the low water months there would be no water coming over the weir, and the sluice gate would be closed to guide water in a different direction. The boys' faces are dark because the sun is almost overhead

arrived too late for thousands of people. Sadly, this could happen again in many countries. Perhaps the lessons to be learned are:
• Droughts recur: governments need to plan to reduce their impact.
• A balance must be kept between water supply and population growth.
• Natural disasters are made worse by wars.

There is more in Part 3 about efforts being made to deal with drought and desertification, and about the need for peace.

Southern Africa

Chapter 14 The front-line states

> **Key words**
>
> Land-locked, SADCC, imbalance, spatial relationships and patterns

The countries of southern Africa are grouped together in this section partly because of their location but also because they have come together politically as the 'front-line' states. Their association began during Zimbabwe's independence struggle, and continues because of their position at the interface, that is the point of confrontation, with the Republic of South Africa. They all need to reduce their economic dependence on the RSA and 9 countries have joined together in SADCC, the Southern African Development Coordination Conference.

Seven things dominate the geography of southern Africa.

- Southern Africa is an area of outstanding mineral wealth. It has the richest mineral deposits of the African continent, and is one of the richest mineralised zones in the world. It has good deposits of 'strategic' minerals, that is, minerals for use in weapons, for example cobalt and uranium.
- This mineral wealth triggered the nineteenth-century European 'grab for land' a century ago.
- Cecil Rhodes, a pioneer coloniser and businessman, hoped to control mineral resources and to 'colour the map of Africa red' for the British Empire from the Cape of Good Hope to Cairo ('Rhodes' Dream'). He meant to achieve this through building railways.
- Mineral sites were often the key to the colonial transport pattern and the division of the land between black and white. Wealth from diamonds (Kimberley, 1865) and gold (Witwatersrand, 1886) financed the railways and decided the routes they took (see Figure 14.3).
- In many African countries before independence, white farmers owned more of the land with better soils and potential. This resulted in imbalance: relatively small numbers of whites had most of the better land; larger numbers of blacks lived on less good quality land.
- There are 4 major land-locked countries in southern Africa: Zambia, Malawi, Zimbabwe, and Botswana. They have no coastlines, so lack direct access to sea routes. They depend on seaports in other countries, including the Republic of South Africa, for all their imports, including fuel oil, and for most of their exports to earn foreign exchange.
- The separate development policy of the Republic of South Africa is the cause of the rift between it and the front-line states. Zimbabwe, Botswana, Angola, Zambia, Tanzania, Malawi, Mozambique, Swaziland and Lesotho have formed the Southern Africa Development Coordination Conference.

The rest of this chapter centres on:
- the remarkable mineral wealth of southern Africa
- export routes
- the Beira corridor

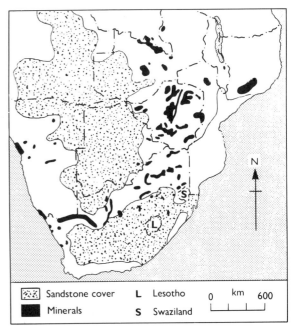

Figure 14.1 Mineral sites in the Basement rocks of southern Africa

- and SADCC, southern Africa's own initiative towards economic and political stability

The mineral bearing rocks of southern Africa

The ancient 'Basement' rocks of Africa are rich in minerals of all kinds. But in much of southern Africa the Basement is covered by younger sandstones, hiding the mineral zones. So the important mining sites are found where the sandstones are missing, in a broad zone running from south-west to north-east as far as Shaba in Zaire.

Use Figure 14.1 to check the following:
1. The stippled area (sandstones) and the mineral deposits marked in black.
2. There are two stippled areas, one on the west reaching to the equator; the second in the south reaching as far north as the latitude of Johannesburg and Swaziland.
3. The sandstones form huge cappings on top of the older 'Basement' beds (see page 121).
4. Where they cover known mineral deposits they make access for mining difficult.

The network of railways and export routes

Use Figure 14.3 to check the main exit routes to ports in southern Africa, particularly for the landlocked countries.
Remember that:

1. the Tazara route and the port of Dar es Salaam are congested and slow
2. the Benguela railway to Lobito has been closed since 1976
3. the Beira corridor railway through Mozambique is constantly under attack
4. the Malawi–Nacala and Limpopo (Zimbabwe–Maputo) lines have also been sabotaged
5. other land export routes are through the Republic of South Africa
6. the map is at a very small scale: the Zambian copperbelt is over 1,500 km from Dar es Salaam, 2,400 km from Lobito and nearly 2,300 km by rail from Beira.

The Beira Corridor lifeline: some problems

Survival for Zimbabwe depends on access through Mozambique to the sea. Zimbabwe spends millions of dollars and has sent thousands of troops to keep the line open. The railway is over 300 km long. A road and an oil pipeline run parallel to the railway. When the oil flow was cut there was chaos for weeks in Zimbabwe.

An interruption in any of these links causes shortages and makes ordinary repairs and maintenance impossible. For example Dunlop in Zambia got no rubber for tyres. The road transport services said: 'We have 350 buses and trucks but only 120 are working. The rest have no tyres or inner tubes and this is just when we need to shift the maize crop.' Usually 15 million

Figure 14.2 New road and bridge, Zimbabwe. The old road and bridge are shown on the right

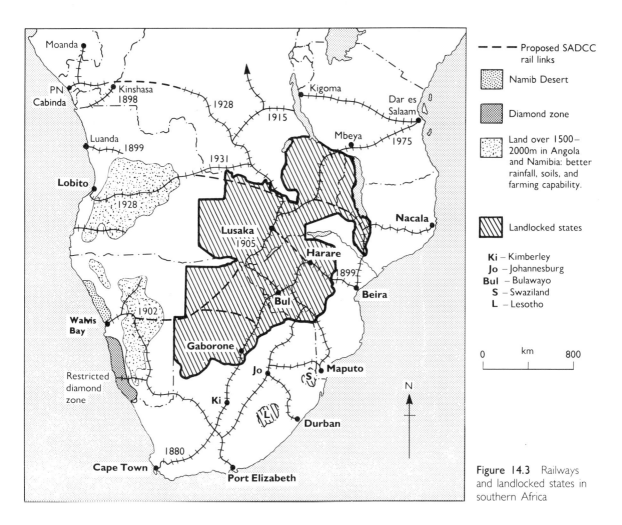

Figure 14.3 Railways and landlocked states in southern Africa

tonnes of goods a year are moved by road. Now it is down to 2·5 million. Refrigerated goods perish in hold-ups.

So what actions are the front-line states taking?

The Southern African Development Coordination Conference was set up in 1979.
• The main aim is to reduce economic dependence on the Republic of South Africa.
• Combined, the countries form a stronger negotiating group.
• They can argue out issues that might otherwise divide them, so are a force for peace.
• They intend that 'outsiders' (whether from the West, Soviet Russia or China) should not become too powerful.
• They work economically but efficiently from Gaborone, Botswana. Their flexibility allows shifts of policy when necessary.

The member countries are taking positive action.
• They have built direct telecommunications links. Previously they had to connect Harare to Lusaka via London, Johannesburg or Cape Town.
• Priority projects are the rehabilitation and reopening of the Malawi–Nacala railway, repairs to the Limpopo railway and new locomotives for the Tazara railway, as well as improvements to the Beira railway.
• In good years food stocks will be stored. There will be joint agricultural research, for example, into drought resistant crops, and 'early warning' given of drought conditions.
• International aid has been gained to restore rail and port facilities.

It is not as straightforward as this sounds. But all these actions are significantly changing the geography of southern Africa and the spatial relationships or links that the countries have with each other, with the Republic of South Africa, and with the outside world.

This chapter has set out some of the problems common to all the countries of southern Africa. Chapter 15 uses Zambian copper mining to show how a country can be put at risk by these, when combined with another world situation: the drop in the value of copper at the beginning of the 1970s. Chapter 16 on Zimbabwe centres on 3 things:
- the land development gap resulting from the colonial period
- the division of agricultural development into 'communal' and 'commercial' sectors
- mineral wealth and railway routes for exports.

Summaries of the other countries of southern Africa are included in this chapter but Lesotho and Swaziland are included in Chapter 17.

Malawi

Capital: Lilongwe, 159,000 (1983)

Exports $242 million
Imports $214 million

Visible trade balance $28 million surplus (1982)

Export commodities	%	Export partners	%
Tobacco	52	West Germany	37
Tea	27	UK	31
Sugar	7	USA	9
Beans & peas	1	South Africa	7

- Malawi has spectacular scenery and a range of climates that encourages diversification and a high population density:
  - the shores of Lake Malawi (450 m above sea level) have 750–1,000 mm of hot season rain
  - the mountains rise to 2,400 m with rainfalls of 2,000–2,500 mm a year
  - the rest is mainly 1,200 m plateau
- Agricultural exports dominate and there is continued agricultural improvement and land registration
- Previously most development had taken place in the south of the country but the new capital at Lilongwe may help to create a new growth point and open up the north and central provinces.
- Zomba, the old capital, has become the main university centre while Blantyre remains the main commercial centre.
- Malawi is landlocked and therefore dependent on rail and road routes through other countries. New hotels have been built to encourage tourism (see page 190).

Botswana

Capital: Gaborone, 60,000 (1981)

Exports $674 million
Imports $555 million

Visible trade balance $119 million surplus (1984)

Export commodities	%	Export partners	%
Diamonds	72	Europe (mainly Netherlands & Belgium)	70
Copper/nickel	8		
Meat	7		
Textiles	5	South Africa	8
Hides	1	USA	7
		UK	4

- Physical conditions in Botswana are some of the most difficult in Africa. The landscape includes rolling plateaus with interior drainage, the Okavango Swamp, and the Kalahari Desert. Rainfall everywhere is sparse and unreliable (see Figure 14.4).
- About 80 per cent of the people live in the higher south-eastern part of the country.
- Since independence in 1966 there have been remarkable changes. Formerly the country was administered from Mafikeng *inside* South Africa, and used the South African rand as currency until 1976. The economy and exports were based on traditional cattle keeping, and on remittances from South Africa where 20 per cent of Botswana men worked at some time in their lives.
- Now the country has its own independent capital at Gaborone. The discovery of minerals, especially diamonds, has made Botswana one of the top ten countries in Africa for income per head.
- The diamond deposits at Orapa may be the second largest in the world. They came into production in 1971 and produce mainly industrial diamonds. Jwaneng, 80 miles west of Gaborone, may become even more important. In 1984, Botswana became the world's largest diamond producer, with diamonds accounting for well over half of the GNP. Other valuable minerals include copper and nickel at Selebi-Phikwe and Matsitama, coal at Moropule, and uranium, manganese, and asbestos.
- Animal husbandry and farming continue to employ about 80 per cent of the working population. Commercial ranching provides increasingly important meat exports, based on a number of new meat packing factories.
- Water for all purposes is a great problem (see Figure 14.5). The government is working on a National Conservation Strategy which will aim to achieve a more efficient use and re-use of water, to control grazing so as to protect soils and woodlands from depletion and to use coal for cooking and heating, instead of wood and charcoal.

The front-line states 125

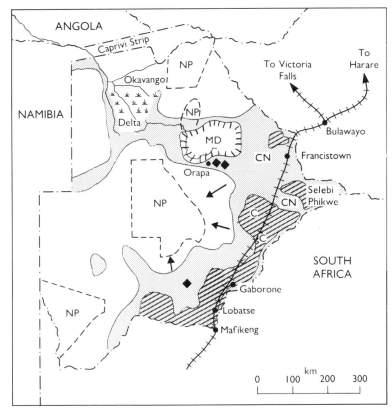

Figure 14.4 Botswana. Most of the western part of the country is covered by Kalahari sands. In the east the soils are better for agriculture and some minerals can be mined from the exposed Basement rocks. The railway also runs through this area

Angola		Capital: Luanda, 1.2 m (1982)	
Exports $704 million		Visible trade balance	
Imports $797 million		$93 million deficit (1984)	
Export commodities	%	Export partners	%
Petroleum	74	USA	33
Coffee	14	Portugal	24
Diamonds	11	Netherlands	12
		West Germany	11
		Japan	10
		France	10

Mozambique		Capital: Maputo, 903,000 (1984)	
Exports $145 million		Visible trade balance	
Imports $711 million		$566 million deficit (1983)	
Export commodities	%	Export partners	%
Shellfish	24	USA	25
Petroleum	17	Portugal	15
Cotton	13	UK	6
Cashew nuts	12	South Africa	6
Tea	11	Netherlands	5
Sugar	7	Japan	5

- Potentially one of the richest countries in Africa because of valuable deposits of oil offshore from Cabinda diamonds in the north-east and good conditions for coffee production.
- The country has suffered from continuing disruption since independence.
- The southern part of the country is partially under the control of the South African backed UNITA movement.
- The Benguela railway has been effectively closed for over 10 years.
- Oil and diamond prices have fallen and coffee production declined.

- Mozambique stretches through 12° of latitude (nearly 2,400 km) which has advantages in the range of crops that can be grown and disadvantages for unity and communications.
- It has a high proportion of coastal plain. The agricultural potential is good, and the Cabora Bassa dam on the Zambezi river could provide cheap power, flood control and irrigation.
- Various valuable minerals include oil.
- An extensive guerrilla campaign has disrupted much of the country.
- Economic production has declined dramatically, and basic food production has been affected by serious drought.

Southern Africa

Namibia	*Capital:* Windhoek, 104,00 (1983)
Exports $386 million Imports $390 million	Visible trade balance $4 million deficit (1984)

Export commodities: Diamonds, Uranium, Copper, Lead, Zinc.

- Namibia is a country of vast potential, but much depends on the outcome of the efforts to establish political independence from the Republic of South Africa.
- The country so far has been run on racially divided lines, similar to apartheid. White ranchers control the best of the farmland, while the black and coloured population (93 per cent) cope with much more difficult conditions where drought and overgrazing are frequent. Only 12 per cent of the country's wealth is available to the non-white population.
- Diamonds are Namibia's biggest asset. They are found in beach deposits on ancient shorelines in an area called the Sperregebiet (see Figure 14.3). Deposits were estimated to last at least until the year 2005 but they have been overmined since 1971 and production is being scaled down for a 1992 closure.
- The trade statistics are incorporated with those of South Africa, as part of the Southern Africa Customs Union. Hence it is difficult to find data for Namibia alone.
  - diamonds are the main export
  - the largest uranium mine in the world is at Rossing and uses power from the Ruacana Falls dam on the Cunene River which forms the boundary between Namibia and Angola
  - fishing offshore has been affected by over-exploitation
  - karakul wool is exported for the fashion knitwear market.

Figure 14.5 This photograph of a water pump in Kenya represents the type of project undertaken by all dry zone countries. The government of Botswana has tried to install water points (taps) in all villages.

Chapter 15 Zambia and the copperbelt

> **Key words**
>
> Land-locked, single commodity economy, diversification, underground mining

Zambia is typical of many African countries, especially the land-locked ones. There were high hopes at independence in 1964 but 'take-off' was based on good copper prices rather than on agriculture. Since then falling copper prices on world markets, and export routes blocked by warring states, have all but wrecked the Zambian economy. 'The economic cards have been stacked against Zambia since independence.'

Thus Zambia illustrates:
- the problems of an economy based on a single commodity, copper (copper still provides about 88 per cent of the exports, see trade summary)
- a country without direct access to the sea
- an economic imbalance from the colonial past with sharp contrasts between urban centres such as Lusaka or the copperbelt, and the rural areas. (There is a study of Lusaka housing on page 187, Part 3.)

Zambia	*Capital:* Lusaka, 538,000 (1980)
Exports $914 million	Visible trade balance
Imports $612 million	$302 million surplus (1984)

Export commodities	%	Export partners	%
Copper	88	Japan	19
Zinc	3	UK	8
Cobalt	3	USA	7
Lead	1	West Germany	3

The Zambian copperbelt

The Zambian copperbelt is part of a huge mining zone stretching from Kabwe (Broken Hill) in the south, to Kolwezi in Shaba (Zaire). The combined resources of the Zaire Koperzone and Zambia total about a quarter of the world's reserves. Together they produce 13 per cent of the world's copper (Zambia's share is 7 per cent) and Zambia is the world's fifth largest producer. In 1983 Zambia exported metallic copper worth $US 800 million. The copperbelt towns, Ndola, Kitwe, Chingola, Luanshya are a group of mining centres spread at intervals for 150 km through the Zambian copperbelt.

Copper mining at Nkana–Kitwe: a study of a large mining enterprise

Nkana mine and the town of Kitwe form an economic island in the heart of rural Africa. One mine and one town might involve over 12,000 workers. Some have their wives and children with them. Some come from villages where the land is farmed mainly by women, old people and children.

This study is in two parts. One is the survey map extract and box which includes an aerial photograph (pages 130–131). You can look at that first if you wish, and answer the questions, and then continue with the second part, below.

Detective work on Figure 15.1.
1. Find:
- the Kafue River flowing from north to south
- the two outcrops of the copper ore body marked
2. Check these *man-made landscapes*:
- The built-up area of Kitwe. It occupies the wedge of land between the Kafue River and the eastern ridge of the copper ore body.
- The Central Business District.
3. Find Nkana mine and name all the features that relate to mining. This is a *mining landscape*. Don't forget transport – 'no exports, no customers, no income'.
4. Now check these features of the *physical landscape*:
- The Kafue River flows north–south.
- Tributaries from the west flow through the town. Find the one that gives the town its name.
- What do you notice about the upper course of the Kitwe stream? What does it tell you about the land between the outcrops of the ore body?
- Is the unusual stream network repeated in the case of the other tributaries?
- What now occupies this position on the Mindola Stream?

128 Southern Africa

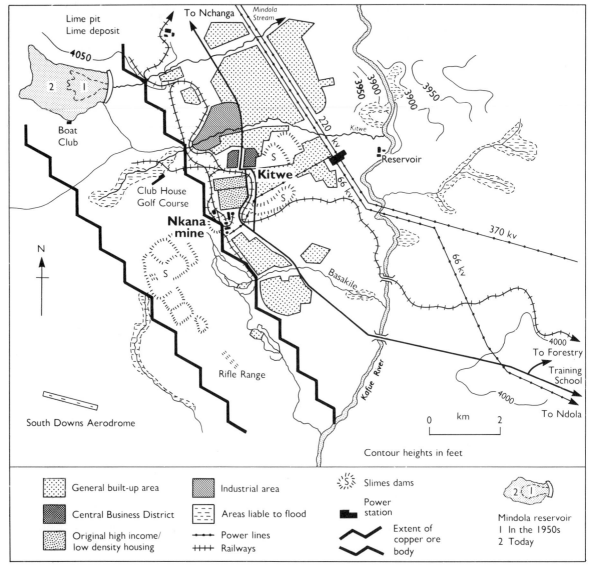

Figure 15.1 Kitwe: a sketch map showing town and mining installations

- Check the symbol for 'liable to flood' in the map key.
- Find out how high the land in this area is. Try to describe it in one or two words.
- Convert altitudes in feet to metres (divide by 3.28).
5 Compare the sketch map, Figure 15.1 with the survey map extract, Figure M7, page 131.

Mining and processing copper ore
Figure 15.2 shows the copper ore body, shaped like a boat. Its direction (or trend) is roughly north-west to south-east.

1 The copper bearing bed is shown in 2 places, one on the west, one on the east.
2 On the west it outcrops at the surface and can be quarried from an open pit.
3 In the east at Nkana, the best copper bed lies below the surface and is nearly vertical. There it must be mined from shafts.

Nkana Mine
Copper was first produced at Nkana in 1939. At that time coal was brought from Wankie (Hwange) south of the Zambezi, a distance of 900 km.

Zambia and the copperbelt

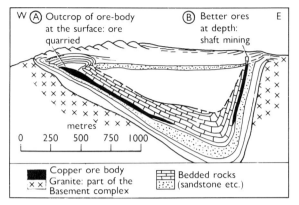

Figure 15.2 A cross section of the copper ore body at Kitwe

During the 1950s electric power came from the Belgian Congo (Zaire); in 1960 the Kariba South Bank power station began to transmit to the copperbelt. Now Zambia has its own Kariba North Bank and Kafue power stations, the latter opened in 1972.

The processes which extract copper from the copper bearing rock are shown in the flow diagram, Figure 15.3.

Labour on the copperbelt

At one time the African mine employees on the Zambian copperbelt were drawn from the country itself, and from Tanzania and Malawi. All 'incomers' lived in the mining townships. So the copperbelt towns were really experiments in community living. Incomers help their home areas because they take or send money and ideas back to hundreds of villages.

In Zambia there are 4 main languages, Nyanja, Bemba, Lozi and Tonga, although there are about 73 different tribes. English is now the medium of instruction in all schools so it can be used in the mines and in commerce and industry in general.

The last 25 years have seen many changes. Some of the old timers have risen to responsible posts. The Chairman of ZIMCO, a Zambian, is top man in the mining industry. Many Zambians occupy senior positions as general managers. Others are Mine Captains and shift bosses.

Junior (middle) and senior staff are 'management' and are not MUZ members. The Mine Workers Union of Zambia (MUZ) negotiates conditions and demands with management, for example, pension schemes, days of leave. This 'ladder' of promotion and change is described because it is typical of what is happening throughout Africa.

Power supplies in Zambia

Zambia's power supplies were originally based on Hwange (Zimbabwe) coal and electricity from Zaire. Later electricity was supplied from the Kariba dam. However, the ups and downs of dependence on other countries led to Zambia's developing the Kafue Gorge power station, Kariba North Bank power station and Maamba coal deposits near Lake Kariba. The Kafue Gorge complex includes a dam, underground power house and a storage and regulation dam at Itcshi Teshi. The reservoir can be used for fisheries and

continues on page 134

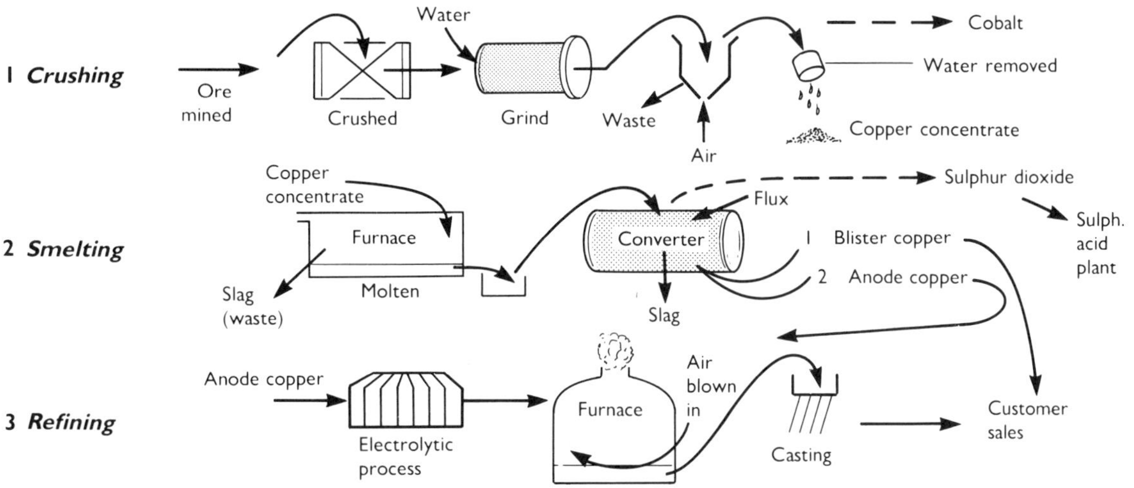

Figure 15.3 A flow diagram to show how copper is produced from copper ore. Similar processes are used for other minerals.

Map extract box 3: the landscapes of a mining town: Nkana–Kitwe

Nkana–Kitwe has special landscapes because it is a town whose main industry is mining.

Look at Figure M7, an extract from Sheet No 1228C3, Zambia, 1986.

1 Find the Nkana mining and processing complex.

2 Find the slimes deposits. They are the waste or spoil tips. There is an immense amount of waste rock and gravel left over when copper is processed.

- The older tips are to the east of Kitwe centre and push residential areas further out, causing the town to sprawl over a wide area.
- The newer tips are to the west between the outcrops. Find reasons why this is likely to be a better site. Conveyor belts are used to move waste to the slimes deposits.

3 Check the following:

- the power station near Nkana mine
- 220KV and 66KV power lines near the eastern edge of the map.

4 Note the amount of space taken up in the town by railway sidings and marshalling yards.

The photograph, Figure M6, was taken from a point on the map near the last 'A' of NKANA, looking in a southerly direction. If you have 2 books, turn one of them around so that the map extract is *oriented* (directed) the same way.

Now use the photograph, Figure M6, to identify:

- the power station
- the ore handling conveyors and large processing buildings, the housing areas and sports grounds

1 Can you tell which is the senior management housing and which is high density housing for manual workers?

2 From which direction was the wind blowing when the photograph was taken?

Figure M6 An air view of Nkana mine complex, looking south

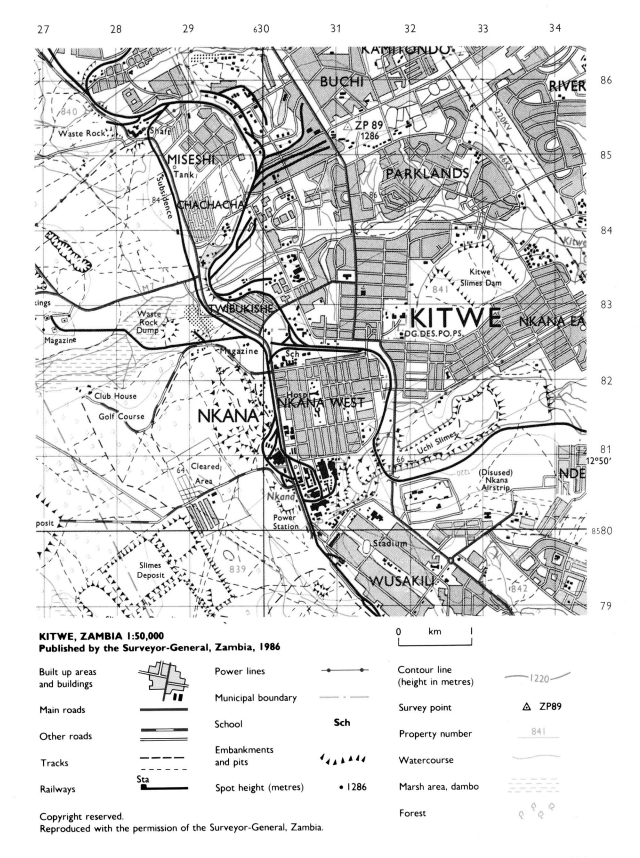

Figure M7 Nkana-Kitwe, Zambia, 1:50 000

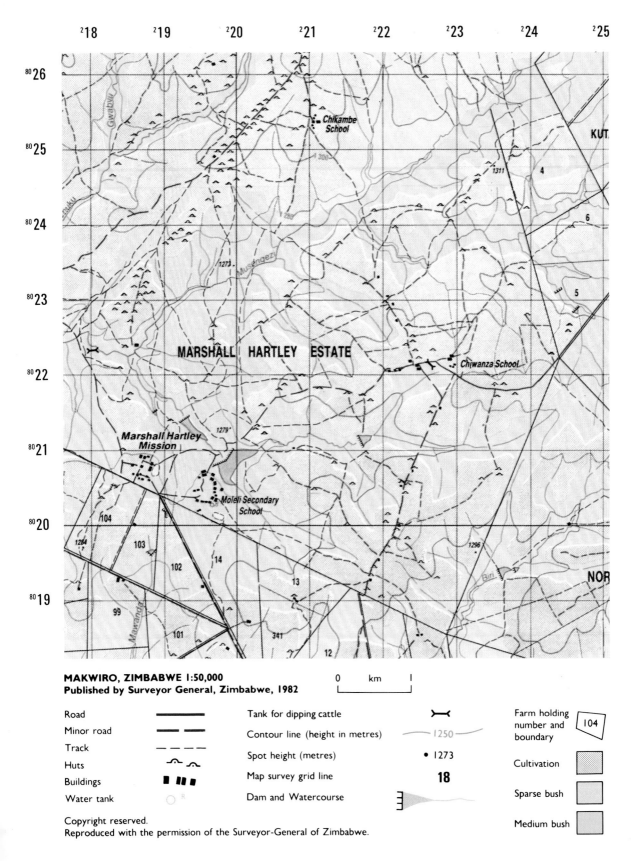

Figure M8 Makwiro, Zimbabwe, 1:50 000

Zambia and the copperbelt

Map extract box 4:
land tenure and farming contrasts in Zimbabwe

Figure M9 is a sketch diagram of the whole map sheet no 1730C4, Zimbabwe, 1982 from which the map extract, Figure M8 is taken. The rectangle near the centre of the diagram marks the approximate position of the map extract.

Look at the sketch diagram and find:
1 The three zones occupied by:
* African villages and communal land
* small commercial farms
* large estates

2 The railway in the south-east; it is about 60 km to Harare.
3 The position of the Great Dyke mineral zone, running from north to south. It was one of the reasons for building the railway, and there are mine excavations marked along the line of the dyke.

Now use the map extract, Figure M8, to find answers to the following questions.
1 Find examples of the symbols shown in the key so that you learn the map language.
2 Use spot heights to find out the actual and average height of the land above sea level in this part of Zimbabwe. How would you describe (a) the landscape and (b) the land use of this area?
3 The Hunyani River lies to the north of the area covered by the map extract. Do most of the small rivers shown on the extract look as though they are flowing in the right direction to join it?
4 The pattern of village settlements is very distinctive. Are the homes and roads in the valleys or along the ridges? Where are the cultivated areas? Why have these locations been chosen.
5 Identify some of the small farms to the south of the Marshall Hartley Mission. Work out the area of a rectangular one, farm 104, and use the scale to measure the 2 sides before multiplying them. (There are 10,000 sq m in a hectare.)

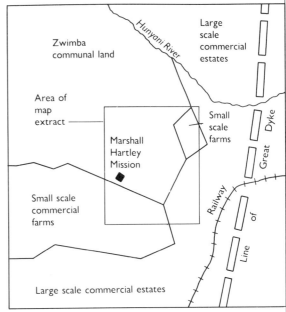

Figure M9 A sketch map of the whole Makwiro survey map

6 How can you tell that the small lakes near the Moleli secondary school are artificial? What are they for?
7 **Note** on the map extract the large scale commercial estates are only represented by the Northwood estate. Find **NOR** in the south east corner. Its full name could not be included because of the small size of the map extract.
8 You can use the survey map grid lines to locate a place to the nearest 100m by giving a grid reference. For example, Chikambe school is at 211 254. First, read the numbers along the top of the map from west to east (the easting). The school is 100m east of line 21. Next, read the numbers on the side of the map from south to north (the northing). The school is 400m north of line 25.

134 Southern Africa

continued from page 129
irrigation water for Kafue Flats sugar.

Maamba coal is not as extensive as the deposits at Hwange and has more ash. Apart from its customary use for power generation in industry, low quality coal and coal dust could become one of the most important contributors to cheap domestic fuel and conservation of woodland throughout central Africa.

Low-grade coal can be used for cooking and heating in the following way.
Look at Figure 15.4.
All you need is
- **coal dust**
- **a handmade metal frame**
- **damp clay**
- **a suitable stove for burning coal bricks**

Coal bricks would be especially valuable where fuel wood is running out and bush clearing is causing soil erosion.

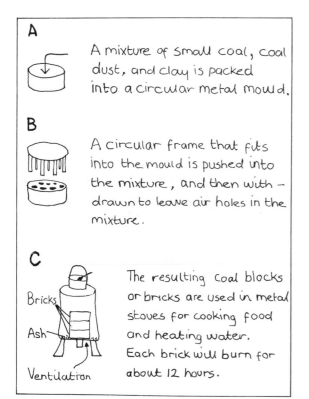

Figure 15.4 Saving the woodlands: coal dust made into blocks for cooking

The future for agriculture

Zambia is a large country (753,000 sq km), and the population is only 6 million. But only one-third of the country is suitable for agriculture, and two-thirds of the population live there.

Good land suited to commercial farming is severely limited. Most of it lies near the railway line (see Figure 15.5, a summary sketch map of the whole country).

The problems are:
- poor sandy soils, some laterite (ironstone) layers, bare granite domes or pavements especially in the southern and western provinces
- a dry season from August to December, in some places even longer
- more than half the country does have adequate rainfall for half the year (December–April), but these are areas where tsetse flies flourish (about 40 per cent of the total area). The bush can be cleared of tsetse flies but they return if the area is thinly peopled
- because the rainfall is adequate, most of the country is covered with bush, woodland, or forest. Much is high woodland savanna with tall trees, difficult to clear.

The land system called *chitemene* has been abandoned. It was very extravagant – a family of 5–6 people might cut and burn 65 piles of brushwood to fertilise their own small cropped area. This was exhausted after 3 years, and had to rested for up to 25 years. As the population increased, more and more brushwood was cut, leading to deforestation and soil erosion.

Zambia now encourages agriculture and *diversification*. Vacant or undeveloped land reverts to the state. It is recognised that agricultural production in rural areas must be improved to:
- raise the living standards of the two-thirds of the people who live there
- reduce the numbers of people flocking to the towns
- supply vegetables, foods, and services to the mining communities and other towns
- encourage small stores and rural industries and services, if farmers have money to spend
- encourage the growth of informal markets and small towns which can provide health, education, and other services
- reduce the amount that Zambia spends on imported food.

Zambia and the copperbelt

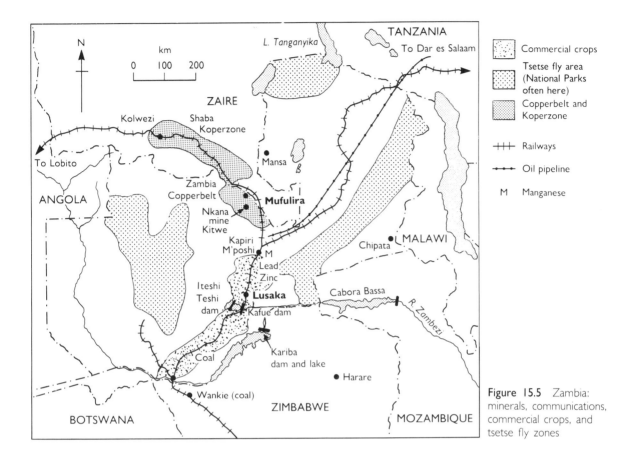

Figure 15.5 Zambia: minerals, communications, commercial crops, and tsetse fly zones

Different farm systems

Much of the farming is still *small scale* and domestic, but there are improved methods, cooperative buying of seeds and fertiliser and marketing, and increased production of some export crops. Figure 15.6 shows different scales of farming in Zambia.

The number of white settler farmers, producing poultry, milk, beef, wheat and tobacco, has decreased. Land vacated by white farmers is now farmed communally. There are some family farm resettlement schemes, which have been very successful. However, in some areas, difficult farming conditions discourage settlement, and the tsetse flies return to thinly peopled areas.

Thousands of small farms produce maize but not enough to meet demands. Zambia grows its own sugar on the Kafue Flats. Now there is a drive to increase yields and to make Zambia totally self-supporting in food, especially cooking oils, maize and wheat.

Commercial stock raising has increased. The government has formed some very large state ranches, between 15,000 and 40,000 ha. Grazing in 400 ha paddocks is controlled and rotated in sequence according to the season, to avoid overgrazing. Water is supplied to each paddock from wells, water troughs, dams or wind pumped boreholes.

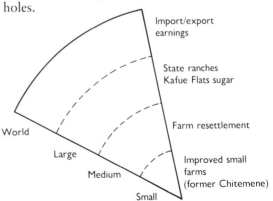

Figure 15.6 Agricultural activities at different scales in Zambia

Figure 15.7 Spraying groundnuts in Zambia

The state ranches cross-breed local cattle such as the Angoni with the humped Boran or Zebu. This produces stock able to use poor grazing, survive high temperatures and irregular water supplies, and withstand diseases. Breeding heifers go to other ranches and farms. There are also fattening schemes. Small farmers fatten cattle on loan from other farms along with their own. They are paid the difference between the loan price and the eventual sale price. Dairy cattle are also raised near urban centres.

Much of this forms the basis for local industry: slaughtering and cold storage for meat, leather working to make boots, shoes, and sandals from local hides.

Willingness to make changes

Zambian society like all African society is dynamic and changing with the times. The good things in a way of life are strongly held, while the bad things are removed. Some African systems are a good response to local conditions. If a system works well people are unwilling to make changes. But some retired workers come back from urban areas and the mines to farm, combining traditional with modern methods.

Exports and transport

Zambia has all the problems of a land-locked state without direct access to the sea for export and imports. Look back at the map of the railway system (Figure 14.3) and remind yourself of the different export routes available. Zambia cannot survive without exporting yet all normal routes have been disrupted in recent years. Through no fault of its own Zambia has suffered acutely from:
- the drop in copper prices
- its position in the middle of the continent

Chapter 16 Zimbabwe: land sharing, farming, and minerals

Key words

Great Dyke, imbalance, land apportionment, dual economy, opencast mining

Two aspects of the geography of Zimbabwe stand out above all others:
- Zimbabwe is potentially one of the richest countries in Africa.
- The colonial period left a legacy of division into black and white land, with two agricultural systems and a dual economy, which Zimbabwe must work hard to bind together.

The major concern of the Zimbabwean government is to reduce the imbalance in land pressure and agricultural productivity between the sharply contrasted parts of the country. So this chapter focuses first on the reasons for the contrasts in different parts of the country which are part of the present geographical reality. Some of the factors are natural, such as soils, climate, and minerals. Others, such as the apportionment of land between black and white, were man-made.

1. Look at Figure 16.1A, which is a map showing the area of high agricultural potential. This area in the centre of the country is almost the same as the area of the highest land – the High Veld (average 1300 m). It has better soils and is wetter and cooler than the Middle and Low Veld on either side to the north-west and south-east.
2. Figure 16.1B shows that the Great Dyke also runs through this area; how the gold belts relate to it; and the way that the railways were built to serve these economically productive zones.
3. Figure 16.1C, is a simplified map showing how the land was 'apportioned' between black and white during the period of colonial rule.

- The division of land between the whites and blacks was fixed by the Land Apportionment Act of 1930. It set aside 51 per cent of the country for white settlement, and 30 per cent for the Africans who were then 90 per cent of the population. The rest was state land. By 1951, the average white farm extended to 1,500 ha, of which only about 10 per cent was actually cropped. This broad land division was continued

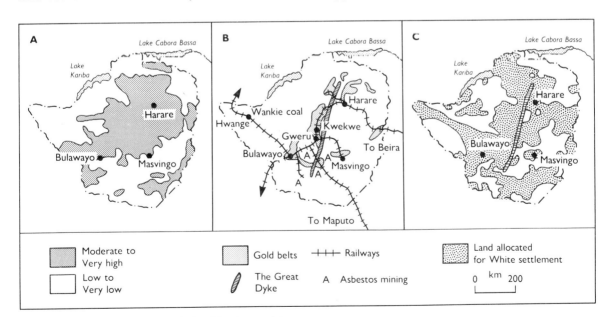

Figure 16.1 Zimbabwe: A Agricultural land potential
B Railways and minerals
C Land apportionment during the colonial period

in the Land Tenure Act of 1969. This gave about 47 per cent of the land to 5 million blacks and about 46 per cent to 200,000 whites. The remaining 7 per cent was national land.
- In the 1950s the whites got the areas of high agricultural potential (Figure 16.1A), plus the mineral sites (Figure 16.1B), while the blacks got the peripheral areas usually with the marginal rainfall and poor soils.
- The result is *two quite different farm systems* and a *dual economy*.
    - commercial farms on over 14 million ha of mainly good land, supporting 1.5 million people in 1980
    - peasant farms on 16 million ha in the communal areas, to support 4 million people on land of lower potential
- The commercial farming area has the benefit of a network of railways, roads, and powerlines, and commercial centres (i.e. infrastructure) far ahead of anything to be found through most of the communal lands.
- The communal areas had no proper service framework. They suffer from poor water supplies, a lack of marketing agencies, poor agricultural advice services, and difficulties in obtaining money for improvements. Look at Figure 16.2 to see another imbalance – in population structure (for more information on population pyramids see page 168, Part 3). Despite this, they have been increasing their production of maize, cotton, and groundnuts proportionally faster than the commercial areas. Some communities are regulating the number of animals that can be kept, so as to avoid over-grazing, and wild game is being encouraged in some areas to promote tourism.
- Over half of Zimbabwe is unsuitable for cropping for one reason or another. A large area, particularly in the communal areas, is formed from the Basement Complex granite rock which provides the mineral wealth, but produces poor soils. The granite outcrops form 'pavements' over huge areas, sometimes with spectacular tor landscape, such as the Matopo Hills (see Figure 16.3). Find the map extract box, pages 132–133, check these contrasts on the map extract and answer the questions.

Sharing wealth, sharing land

The new government has worked hard to redistribute better land to black farmers. Former white-owned farms have been bought on a 'willing to sell' basis and subdivided to provide

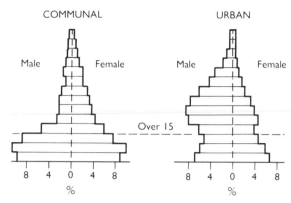

Figure 16.2 The two population pyramids of Zimbabwe show how men in the communal areas have moved to the towns

small holdings suitable for family farming. Some large farms are now operated on a cooperative basis.

In the *communal areas*, traditional farming continues.
- There has been a great effort to improve community services in the rural areas where about 75 per cent of the population of Zimbabwe lives. More water boreholes and taps, clinics, and schools have been provided, with the number of children receiving free primary schooling increasing from 0.8 m at independence in 1980 to 2.5 m in 1985.
- Much more of the important maize crop is now produced in the communal and resettlement areas – well over half the crop in some years. Before independence, the largest total annual maize crop produced by the peasant farmers was 65,000 tonnes. In 1984, it was 380,000 tonnes. This massive increase in production has resulted from favourable market prices, good rainfall, farm training schemes, and improved use of seeds, fertilisers, and pesticides.
- Much of the crop is used for home consumption, but sometimes there is a valuable surplus that can be sold to help other areas where the rains have failed.

The *commercial farming areas* are important because they produce most of the exports that provide income for Zimbabwe to buy things from abroad.

1. Look at the trade summary for Zimbabwe and add up the total percentage of export income that comes from crops, food, and live animals.
2. Now do the same for the 4 mineral based exports that are listed.

Figure 16.3 The Matopo Hills near Bulawayo

Zimbabwe

Capital: Harare, 681,000 (1983)

Exports $1,344 million
Imports $990 million

Visible trade balance $354 million surplus (1984)

Export commodities	%	Export partners	%
Tobacco	19	South Africa	18
Gold	13	UK	13
Food & live animals	11	West Germany	9
Ferro-alloys	10	USA	6
Cotton	8	Japan	5
Asbestos	5	Italy	5
Nickel	4	Botswana	5
Copper	2		

The government faces a dilemma. It wants to spend most money on improvements in the communal areas, such as resettlement and infrastructure projects, and better farming advisory services. But the more important export earnings come from the commercial sector. For example, commercial farms often produce 100 million kilograms of Virginia flue-cured tobacco per year. The small peasant farms produce only about 6 million kilos, and it is lower quality 'burley' tobacco which fetches a lower price.

Tobacco growing in Zimbabwe

Tobacco is a very exacting crop, both in the standards of production and the demands made on the soil.
• Soil fertility is soon lost, so a tobacco crop is only grown one year in 5, in rotation. Thus an estate must have about 200 ha of reserve land in proportion to every 50 ha of tobacco.
• Tobacco is liable to attack by eel worm, so the land must always be kept clean. Land has to be fumigated to reduce eel worm.
• Some grasses resist eel worm, and they can be grown and grazed for at least 3 years. Grass for seed is also a saleable commodity.
• It takes a very heavy investment to produce tobacco: labour, equipment, seeds, fertilisers, insecticides, etc., as well as processing, storage, packing, and transport costs. But the gross income is about 5 times that for maize, and 2 or 3 times that for cotton or groundnuts.
• Annual cash crops can be grown instead of grass, for example maize, but they reduce the soil fertility and can cause soil erosion on sandy soils. Large mixed estates also carry beef or dairy cattle and other stock.

Consider these points and the tobacco fact box on page 140, and work out 3 important reasons why it is difficult for small producers to compete successfully with the large tobacco estates.

Zimbabwe is opening up tobacco farming to small scale growers in 2 ways:
• The government has bought large estates as they come on the market, and has divided them into small farms for resettlement. The large estate organisation converts fairly easily into a cooperative of small farmers who can follow the rotation and other methods established in the large-scale operation.
• Farmers in the communal areas are being encouraged to grow tobacco but because of land pressure it is much more difficult to integrate it

into traditional farm practice and still keep the years of fallow.

Cattle farming is another important commercial enterprise. Beef exports are important, and a special programme has been introduced to provide fencing to reduce foot and mouth disease, and to ensure that EEC food hygiene regulations are met.

> *Facts:* **Tobacco**
>
> *Climate*
> - Temperatures: from 13–27°C
> - Altitude: 900–1,500 m
> - Rainfall: 635 mm in growing season
> - Soils: light, well-drained
>
> *Farm systems*
> - Estates (Malawi and Zimbabwe) and small farms (East Africa)
>
> *Cultivation*
> - Seeds sown June–July in nursery
> - Transplanted September–November on ridges
> - Matures in 6–9 weeks, reaped (leaves picked) from bottom upwards. Tied or clipped in bundles for curing
>
> *Processing*
> - Fresh leaves are cured, i.e. dried under controlled conditions
> - 2 types: flue-cured in barns with air vents heated by coal or wood; air-dried (burley) which fetches lower prices
> - Graded for uniform colour, quality, leaf position on stem, size, etc., before baling
>
> *African production*
> Zimbabwe produces 35% and exports 50% of African tobacco; Malawi 22% and 27%.

Minerals in Zimbabwe

Figure 16.1B shows the Great Dyke running for nearly 500 km through the country. The Great Dyke forms a highly mineralised zone and the related igneous bodies contain a greater variety of minerals than is found in any other single area of Africa. They include asbestos, chrome ore, gold, nickel, iron ore and copper. In addition to this, there are vast deposits of coal at Hwange (Wankie) and other areas. It was the lure of minerals, especially gold, that brought the first white settlers into Zimbabwe. The railway network was built to secure control over the country, and to carry minerals to the coast. Zimbabwe's mineral sites were 'lean' and not on the same scale as those of South Africa but it was the spur to development and the railways and other support services have helped commercial farming to develop.

Mining accounts for between 35–45 per cent of Zimbabwe's export revenue in most years. Asbestos, nickel and copper are important mineral exports and so is gold. Look back at Zimbabwe's export figures. The trade summaries throughout the book are helpful in giving basic figures and information about a country's production. But they have to be used with care because sometimes important *products* are not exported. Coal is not listed but Zimbabwe is the second most important *producer* in Africa.

Coal has played a very important part in the development of the country fuelling railway locomotives and industry both for Zimbabwe and the Copperbelt. The Cape to Cairo railway was re-routed in order to use Wankie coal.

The Wankie colliery near Hwange produces about 3 million tonnes of coal and up to 200,000 tonnes of coke a year and the estimated reserves are 1,300 million tonnes of good quality coal and possibly 6,000 million tonnes of poorer quality coal: enough to meet the countries requirements for several hundred years!

Electricity supplies for Zimbabwe from the power stations at the massive Kariba dam on the River Zambezi are now supplemented by supplies from the thermal generating station at Hwange.

Coal production at Wankie Colliery, Hwange, Zimbabwe

There are 4 main resources for this study:
- A fact box of general information.
- A map of the company's concession area (Figure 16.4).
- A line drawing (Figure 16.5).
- A landscape sketch of Hwange town location (Figure 16.6).

These combine to show that this enterprise can make an outstanding contribution to development in southern Africa.

Wankie coal is remarkable because:

- The coal is produced from one immense seam between 2 m and 14 m thick.
- In the opencast area, this bed, or seam, lies only a short distance (20–30 m) below the land surface. It can be quarried from an *opencast* pit after the cover-rock has been stripped.
- Thus, although some coal is still got by deep-shaft mining at a depth of 60 to 350 m, 80 per cent of Wankie's coal now comes from opencast workings.
- This one great coal seam provides two qualities of coal:
  – *the upper third* produces power station coal (more ash)
  – *the lower two-thirds* produces high grade coking coal

Facts: Wankie Coal: its formation and production

- The coal seam was formed during a tropical wet period 200 *million* years ago.
- Lush vegetation grew in a swamp, rotted down to form peat, and was later covered with sand and gravel. High temperatures and pressure transformed peat into coal.
- First mined underground at colliery No. 1. Railway re-routed and the first train load of coal left Wankie in 1903. Previously vast areas of woodland along the track were cut for fuel.
- Supplied the then Belgian Congo and Northern Rhodesia (1909). No. 2 colliery (1919) began to supply Copperbelt.
- Opencast started in 1970, expanded after No. 2 underground colliery explosion in 1972 and now provides 80 per cent of total production.
- The by-products from coking the coal are gas (which is recycled for fuelling the coke ovens) tar, ammonia, and benzol.
- Tar is especially useful for road surfacing, ammonia is used in processing minerals, benzol is distilled and blended with petrol.

Wankie produces each year:
- more than 3,200,000 tonnes of coal
- about 200,000 tonnes of coke
- about 7,050,000 litres of tar
- nearly 1,500,000 litres of benzol
- more than 1,300,000 litres of ammonia liquor

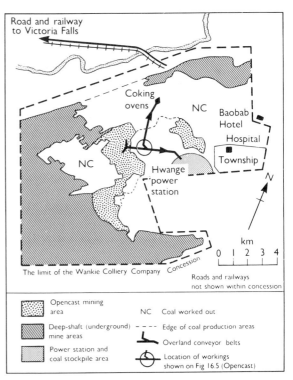

Figure 16.4 Wankie Colliery: mining concession area and coal resources

Now check the following on Figure 16.4
1. The huge size of the Wankie Colliery Company concession. Find the distance from north to south, and east to west.
2. The position of the *underground* coal at the outer edges.
3. The *surface* (opencast) coal nearer the centre.

The two types of coal quarried in the opencast area influence the layout of the colliery complex. One huge conveyor belt takes power station coal to stockpiles at the generating station. A second takes coking coal to the coke ovens. The place where they separate is shown by a small circle.

The opencast mining area at Wankie

Figure 16.5 is a line drawing of the coal landscape. It tries to give a 'bird's eye view' from above the circle marked on Figure 16.4 almost in the middle of the map of the workings.
Start at the back of Figure 16.5 and work towards the front left corner. Make sure you understand what happens at each stage.
1. The landscape of low hills is levelled (*pre-stripped*) (1) so that the huge *dragline* machine

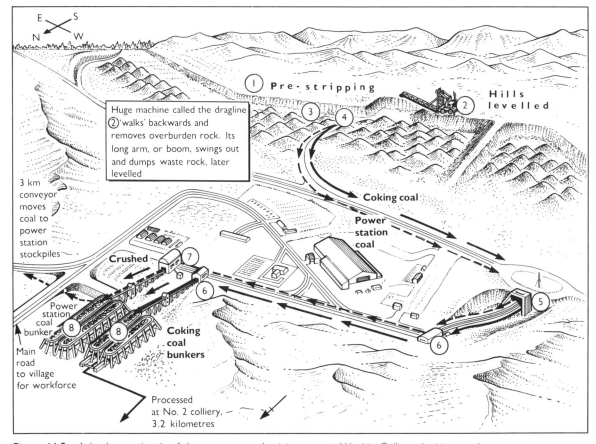

Figure 16.5 A landscape sketch of the opencast coal mining area at Wankie Colliery, looking south

(2) can remove the cover rock (overburden). The heaps are smoothed out later.

2. Power station coal is removed from the upper third (3), coking coal is quarried from the lower two-thirds at (4), and each is taken by truck to the tip (called a grizzley) at (5).

3. At (5) each type of coal is loaded onto moving conveyor belts and travels separately via transfer-houses (6) and (7) to huge storage bunkers (8).

4. At (8) the two types of coal go in different directions, still on moving conveyor belts:

- power station coal travels 3 km east to Hwange Power Station stockpiles
- coking coal is conveyed over 3 km north to the coke processing ovens for coke and by-products production. This is on the site of the old No. 2 pit, where coal is no longer mined.

Figure 16.7 is a line drawing of the Hwange area viewed from the top of the Baobab Hotel.

What happens to Wankie coal

- Power station coal takes priority. It fuels the on-site Zimbabwean Electricity Supply Authority (ZESA) thermal generating station and thermal power stations at Harare, Bulawayo, Umniati and Mutare.

- The coking coal's big customer is ZISCO (Zimbabwe Iron and Steel Company) at Redcliff, about 14 km from Kwekwe. This site has deposits of high grade iron ore, limestone and manganese close together. Only coking coal was missing and could be railed from Hwange, and be coked in ZISCO's own coke ovens. The by-product gas from the ovens *fuel* the plant. Some is now being brought from Buchwa and other centres.

- The third customer is agriculture. Look back to the tobacco fact box to find out why.

- Coal is sent all over Zimbabwe and by rail

Zimbabwe: land sharing, farming, and minerals   143

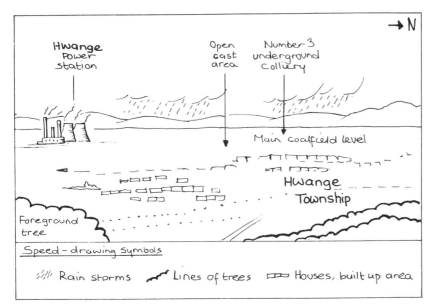

Figure 16.6 A very simple line drawing of the Wankie coal field seen from the Baobab Hotel, Hwange

to SADCC countries. Transport problems limit exports beyond that.
• The Wankie Colliery Company holds a key position in the development of southern and central Africa. The government is trying to promote the use of coal in rural areas, especially where there are high population densities, in order to preserve woodland cover. Coal as such is not easily adapted for use in rural homes (refer to Zambia, page 134). Coal bricks are cheaper, can be used as slow burners both for cooking and room heating in cool season weather. Perhaps someone should experiment?

Zimbabwe past and future

Zimbabwe takes its name from the ancient fortified capital of a civilisation that was based in the south eastern part of the country at least 800 years ago, possibly much older. The impressive ruins that remain show us that this was a well-organised trading state drawing on the resources of a large area. The country regained independence in 1980, following a period of about 100 years of white rule. There was a fierce war of liberation and many whites left when the new black government took over. But many stayed, and some have returned, to share in a well developed multi-racial society.

Figure 16.7 A training session in telecommunications in an older communal area near Harare, Zimbabwe. Note: safety belts clipped to the main telephone post; protective 'hard hats' and spectacles; the white china insulators for telephone wires on the crossbars

144 Southern Africa

The future of the country depends on maintaining a prosperous agriculture and dependable routes for exports. Zimbabwe is land-locked, so all export commodities and vital imports have to travel through other countries. (Check Chapter 14 to see how the countries of SADCC are cooperating to reduce their dependence on South Africa.)

The shortest route from Zimbabwe to a major seaport is the railway, road, and oil pipeline corridor through Mozambique to Beira. The Beira Corridor is of great economic importance but it has been disrupted by sabotage. Zimbabwe has been helping to guard and repair the railway line and it now handles 6 times as much traffic as in 1984. Work should also start on repairing the Limpopo railway line to Maputo (out of operation since 1983) which will take traffic from Bulawayo and the southern parts of Zimbabwe. Reopening these lines depends on a working partnership between governments and private businesses, many of them multinational. International aid has been provided to improve both the railways and the port facilities at Beira.

Harare, capital city

Harare (formerly Salisbury), like many African cities, has older and newer sections. The central area is laid out on a grid plan related to the railway line. The north–south roads are called 1st, 2nd, 3rd Streets, etc. Other street names show political and historic links, for example, Samora Machel, Julius Nyerere; and Livingstone, Stanley and Speke, the explorers.

The line drawing, Figure 16.8, is of the photograph on the back cover. It looks across Cecil (Rhodes) Square to the Anglican cathedral on the corner of 2nd Street and Baker Avenue, with older buildings nearby. Most of the high rise buildings are recent:

- on the left, the shopping district where some areas are closed to traffic.
- in the background, the office blocks of large firms in the CBD.

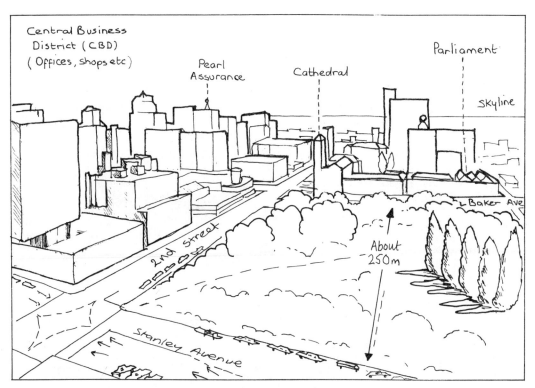

Figure 16.8 A line drawing of the colour photograph of Harare on the back cover. Use the photograph to find the CBD: different building styles related to colonial and modern development; and how one-way traffic is guided at crossroads

South

Chapter 17 The Republic of South Africa

Key words

Apartheid, surplus people, interface, mineral wealth, gold, diversified, industrial development, commercial farming, water resources

Apartheid is the dominant factor in the geography of South Africa. This form of separate development has arisen because of two things:
- the enormous wealth and potential of the country
- the determination of the ruling white group to control the country's resources

The peoples of South Africa are classified by law into four racial groups: white, black, coloured or Asian. Whites occupy 87 per cent of the land area, leaving 13 per cent for blacks and other ethnic groups. So although South Africa is the richest country in Africa in terms of resources and gross national product, access to the wealth of the country is not equal for all groups.

The importance of South Africa: an economic and technological giant

It is impossible to exaggerate the importance of the Republic of South Africa in *southern Africa*. The country has:
- a little under 20 per cent of the land area
- 36 per cent of the population
- but produces about 75 per cent of the wealth

Its position in relation to the *whole continent* of Africa is even more amazing. Look at Figure 17.1. Although the country accounts for only 4 per cent of the area of the continent, and 6 per cent of the total population, it produces 20 per cent of the gross product, and consumes over 40 per cent of the total energy output.

Now look at the trade summary.
1 Does South Africa have a visible trade gap? Check the country summaries for the other countries named on Figure 17.1 and see what their trade balance is like.
2 What commodity earns most foreign exchange?

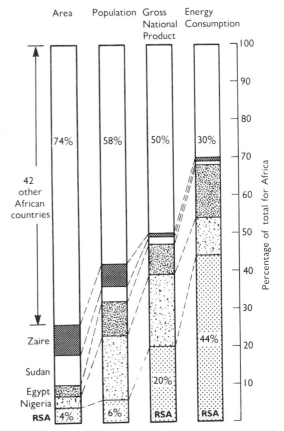

Figure 17.1 The Republic of South Africa compared with other African countries

The Republic of South Africa		*Capital:* Pretoria, 850,000 (1980)	
Exports $8,693 million		Visible trade balance	
Imports $7,614 million		$1,079 million surplus (1984)	

Export commodities	%	Export partners	%
Gold	47	USA	10
Base metals	10	Japan	8
Other precious stones & metals	10	Switzerland	7
		UK	4
Textiles	3	West Germany	4
Chemicals	3	African countries	4

The table shows the importance of minerals in earning foreign exchange as exports. Mining provides about 15 per cent of the gross domestic product of the country with gold accounting for about three-fifths. South Africa is the world's largest producer of gold, gem diamonds, chrome, and vanadium, and is the second most important producer of several other mineral products, including platinum, manganese, and other rare special metals. For many of these items, the Republic of South Africa (RSA) is the main source outside the Communist bloc, so that supplies are of especial importance to the western industrial countries.

But unlike most African countries where the economy depends on a very limited range of resources, South Africa has a strong and well-diversified economy with many other products as well as minerals. There is a highly developed manufacturing sector, contributing about one-third of the total wealth, and an important services sector.

What counts as wealth?

The wealth and development of a country are often measured by its material possessions:
- the average income per head of the population (GNP per capita),
- its consumption of energy,
- the number of people living in towns,
- the length of its roads and railways and the number of cars on the road.

These statistics are also quoted as an index of *level* of development. Such information is a general indicator of how much is being produced. However it often does not take account of the real wealth of home-grown foodstuffs, home-made crafts, and goods and services possessed and circulating, especially in the rural areas of Africa, which never find their way into the statistics.

South Africa has the third highest average income per head in the African continent. This achievement is remarkable because the calculation is based on the large population figure of over 30 million. The two African countries where the income per head is higher (Libya and Gabon) have much smaller populations, and obtain most of their wealth from oil production.

However this average figure hides a very unequal sharing of wealth between the racial groups in South Africa. Black wages are very low, while the minority white population has an exceptionally high average income per head. Their very high standard of living is made possible by a low cost labour force made up of other racial groups.

South Africa has the resources and expertise to become the economic dynamo of the whole of southern Africa. But this cannot happen fully until the government of The Republic of South Africa adopts policies that are acceptable to the black majority of the population, and to the governments of the states to the north.

Minerals in South Africa: 100 years of gold

Most of the minerals of southern Africa are found in the Basement rocks, exposed in a broad zone from Shaba in Zaire to Namibia. In much of southern Africa the Basement rocks are covered by younger sands and sandstones which hide the mineral zones, though deposits of coal are found in some newer rocks (check Figure 14.1, page 121).

The most famous gold-mining area is the Witwatersrand – the Rand. (The rand is also South Africa's unit of currency.) It is a west–east ridge composed of beds of resistant rocks called the Reef. Gold reefs were originally river gravels dropped on the shore of a prehistoric inland sea. They came from the rocks of the ancient Gondwana continent, with pebbles, sand and mud alternating in times of flood and low rainfall. The pebbles were smoothed and rounded just as they are along present-day coasts. The loose pebbles attracted gold particles. Later they became cemented together with finer material and gold

Figure 17.2 A group of apprentices at the de Beers mines school at Kimberley, South Africa

The Republic of South Africa

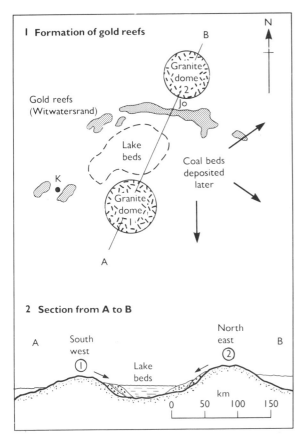

Figure 17.3 The formation of the gold deposits in the Transvaal.
1 An inland sea formed between the original granite domes in the basement rocks. The gold reefs are old shoreline deposits.
Jo: Johannesburg, K: Klerksdorp.
2 A section from A to B shows granite domes, gold reefs (pebble beds) and lake deposits from the former inland sea

was trapped. The whole business of gold mining is that of releasing the gold deposits from the cemented pebble-beds called conglomerates.

The pebble beds follow the old shorelines, so they curve round in a huge arc, 300 km long, from Johannesburg to Klerksdorp and Welkom.

Check the following on Figure 17.3.
1 The general shape and position of the gold arc.
2 The position of the gold-bearing reefs *between* the two granite domes. Materials eroded from the higher granites were washed into the prehistoric inland sea.
3 The section shows how the gold-bearing reefs dip far below the land surface.

Labour in the gold mines

Nearly half a million blacks work in South Africa's gold mines, many of them from other countries. The following is a breakdown of the home area of workers in South Africa's gold, coal, platinum, copper and lead mines in 1985.

South Africa	332,915
Lesotho	108,401
Mozambique	50,885
Botswana	18,830
Malawi	18,652
Swaziland	13,238
Total	542,921

Figure 17.4 shows a goldminer's average monthly wage. The law on job reservation excluded blacks from skilled jobs, but they are now being trained as surveyors, ventilation checkers etc. Increased mechanisation reduces the number of unskilled workers.

A GOLDMINER'S AVERAGE MONTHLY WAGE		
	1971	1984
Whites	R386 (£138)	R1,800 (£643)
Blacks	R18 (£6.50)	R358 (£128)

Source: Chamber of Mines

Despite a big relative improvement in Black wages, the absolute gap between White and Black pay is getting bigger. Blacks get free board and lodging on the mine compounds, which the Chamber values at R120 a month. But White miners also get their houses virtually free, paying only about R15 a month in rent.

Figure 17.4 A goldminer's average wage

Underground (deep shaft) mining to compare with open cast (Wankie)

Use Figure 17.5 to find and check that you understand the following:
Underground
1 The gold reefs dip steeply down from the ridge where Witwatersrand outcrops at the surface (1).

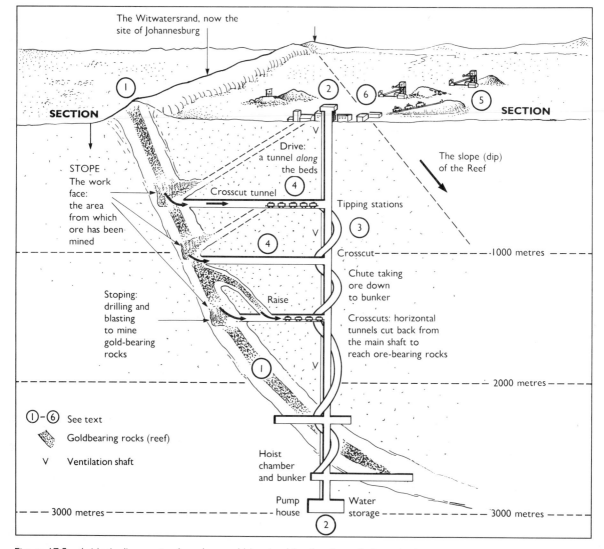

Figure 17.5 A block diagram to show how gold is mined by the deep shaft method

2 Vertical access shafts are used to shift workers, and reef ore. There are also ventilation shafts
3 The main shafts are bypassed by the tipping chute (3) sending ore down to the bunker from the tipping stations. It gets broken up on the way.
4 The cross-cut tunnels are the main access roads from the main shaft to the gold bearing rocks (4). The small railways carry trucks of ore and workers.

Surface
5 The biggest feature covering huge areas is the waste tips (5) (like the slimes dams at Nkana, page 128).

6 The ore is brought to the surface from where it is moved to the crushing station and grinding mill (6) before processing.
7 How deep is the lowest level and what happens there?

Johannesburg is the refining centre for the whole of southern Africa except Zimbabwe. Security measures are strict and armoured cars are used to move the gold around.

In 1984 South Africa produced 48 per cent of the world's gold. The other gold producing countries of Africa are Zimbabwe (12th world producer), and Ghana, Zaire and Zambia.

Check back to the other mining studies, Nkana, pages 127–129, and Wankie, pages 140–143.

The Republic of South Africa 149

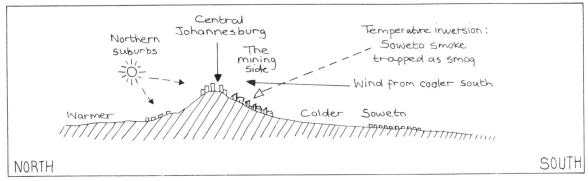

Figure 17.6   A north–south transect across Johannesburg to show how land use relates to weather

Johannesburg: the transformation of a city

It is only just over 100 years since the major discoveries of gold on the Witwatersrand. Now Johannesburg, which grew up there, has the largest concentrations of both whites and non-whites in Africa. In some ways it has become South Africa's racial front line and the interface between the First and Third Worlds.

Johannesburg sums up the story of the last 100 years in southern Africa.
- The site of Johannesburg is a ridge (Witwatersrand) in the great African plateau at about 1,500 m above sea level.
- Southern Africa was open veld (prairie) and much of it was thinly settled and grazed by Africans who moved down from the north.
- In the mid-nineteenth century the Transvaal was established as an Independent Boer Republic.
- The first 'get rich' rush and the surge of railway building from Cape Town to the north began with the discovery of diamonds at Kimberley. Then gold was discovered on the Rand in 1886.
- A simple lay-out was planned for Johannesburg to avoid the chaos of early mining sites like Kimberley.
- As Johannesburg developed into the metropolis of today, land-use zones (functional areas) emerged: mining sites; business and administration; and clusters of residential districts often loosely related to race – white, black, Asian.
- The lay-out is logical (see Figure 17.6). The south side is the mining side. The gold reefs dip steeply towards the south, so the early mines concessions and work camps, waste dumps, and later workers' dormitories occupied the south side of the ridge.
- It is not surprising that Soweto, the largest

Figure 17.7 & 8   Soweto. The first photograph shows how one resident has improved his home. Compare it with the house background right. The second photograph gives a good impression of an improved Soweto neighbourhood

black residential area, (see Figures 17.7 and 17.8), is on the south side and that the smart white residential suburbs are on the north. The climate is also better on the north. There is more smog (smoke-fog) on the south.

150 South Africa

Figure 17.9 Separation of black and white commuters at a railway station, waiting to travel to the centre of Johannesburg

- As gold was worked out mines closed. The white tips, so long a feature of photographs, have been grassed and landscaped. Some are being reworked. But the south is still the less attractive side of town. Old workings also cause subsidence.
- Former mines dormitories have been taken over by the government and used for incoming industrial workers.
- There are changes in central Johannesburg, once an exclusively white area. The expensive shops and some hotels have moved out to the northern suburbs, and central Johannesburg is now an unofficial mixed race area.

> 'Johannesburg is the place where all the races come together by day and go their separate ways at night'

Black workers travel in from the locations on special trains on the Bantu Areas Line, which are 'station to station' so that they can only get off at permitted stops (see Figure 17.9).

Possibly 2 million people live in Soweto 18 km from central Johannesburg, making it one of the largest 'cities' in Africa south of the Sahara. But Soweto was built as a dormitory for migrant workers. It has no real city structure.

There have been changes in Soweto in the last few years. In some areas homes are being bought and improved. There is a lot more free enterprise. *The Sowetan* is a flourishing local newspaper, sold also on Johannesburg main streets. A huge fleet of taxi-vans, take workers into central Johannesburg, avoiding the overcrowded railway.

Industrial development in South Africa

There are only 4 main industrial areas in South Africa, plus a wide scatter of mining and agricultural industries. The latter includes growth areas and so-called border industries.

Area 1. The Pretoria–Witwatersrand–Vereeniging area, known as the PWV triangle, is outstanding and will attract more growth.

Areas 2, 3, and 4 are related to ports: Durban, Cape Town–Table Bay, and Port Elizabeth.

Together these 4 areas produce nearly 80 per cent of South Africa's industrial output and they employ three-quarters of the workers in industry. Figure 17.10 shows these areas and some of the minerals they are based on as well as the importance of railways and ports in the development of mining and industry in South Africa. Figure 17.11 is an enlargement of the Witwatersrand gold fields (the west–east axis) and the new north–south axis through Pretoria, east Witwatersrand and Vereeniging.

The country has an elaborate system of railways and main roads linking the mines, industrial areas and ports. Power and water resources have been intensively developed. Coal makes an important contribution to infrastructure, including oil-from-coal plants, as well as being a major export. The infrastructure services help new industries to start and grow, and play a vital part in the success of the economy.

A new mining and industrial axis in South Africa: the PWV triangle

From the 1880s the focus of development in South Africa was the west–east axis of Witwatersrand. Nearly 100 years later, as the gold axis declined a new axis has developed. It has become known as the PWV (Pretoria–Witwatersrand–Vereeniging) triangle, described as 'the biggest industrial complex between Milan [in Italy] and the Cape'. The north–south axis line crosses the Witwatersrand near Germiston.

The Republic of South Africa 151

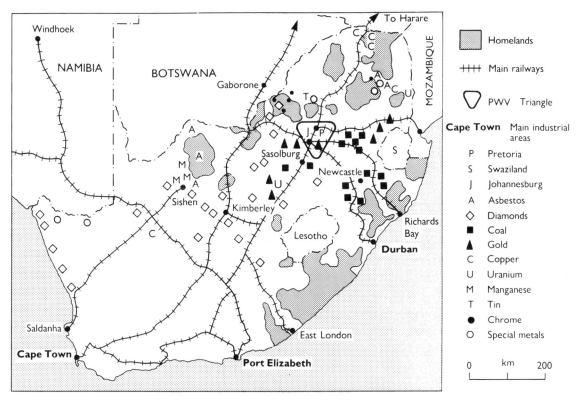

Figure 17.10 South African mineral resources, industrial areas and black homelands. Saldanha and Richard's Bay are both new port developments. They have railways from mining/industrial areas. Richard's Bay has an oil pipeline to Sasolburg and Johannesburg

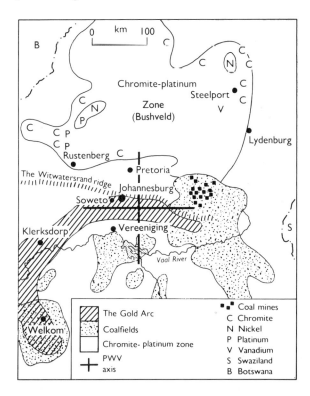

The so-called triangle is a huge metropolitan area about 500 km from north to south and in places nearly as wide from west to east. It is based on some of the richest mineral deposits in the world: coal, platinum and chrome as well as gold.

As early as the 1950s the coal deposits at Sasolburg, south of Vereeniging, distilled oil from coal. Sasolburg, the world's largest oil from coal plant, gets its name from the initials of the South African Coal, Oil, and Gas Corporation. Now there are Sasol 2 and Sasol 3 plants in the Eastern and Northern Transvaal. The iron and steel developments are earlier still: 1928 near Pretoria and 1942 at Vanderbiljpark. All the big state corporations are known by their initials, for example, ISCOR for the iron and steel corporation.

These and other developments have expanded as a result of world situations: the increased price

Figure 17.11 Mineral zones and industrial axes

of mineral oil, and South Africa's move towards self-sufficiency in the face of world talk of sanctions against apartheid. The latter has also had some effect on the volume of exports from South Africa.

Border industries and smaller towns

The latest industrial policy is dispersal. This moves industry out to areas where there is plenty of labour, rather than bringing labour to industry. It fits with the government's policy of growth areas and 'border industries'. The latter, located on the border between white and black areas, serve two needs. They provide jobs for black people who cannot get a living from the land in the homelands and reduce the number of blacks in the towns. Rosslyn is an example. It is on the railway which separates a white from a black area about 20 km north-west of Pretoria.

In South Africa there are hundreds of small towns like Beaufort West, Ladysmith, Calvinia, Nylstroom, Christiana, Theunissen and Franschnoek. The names tell you something about the white South Africans who started them. Each small town has a black township like a satellite about 2 km away. The main road goes through the centre of town and the white shopping area, with a railway running alongside. The black township is away on the other side of the railway track. The area between the two would be a natural choice for light industries needing a black labour force.

South Africa's ports

Ports are crucially important to South Africa. New ports have been built to handle special cargoes, especially Richard's Bay (coal) and Saldanha Bay (iron ore). The large historic ports are Cape Town–Table Bay (see Figure 17.12), Port Elizabeth, East London, and Durban.

Durban is South Africa's main port for general

Figure 17.12 The site of Cape Town, looking south. European settlement at Cape Town dates from about 1650. Find: (1) the historic core area (castle, parliament) now part of the CBD; (2) the newer sections of CBD on reclaimed land on the foreshore; (3) the older Victoria dock; (4) newer docks; (5) Table Mountain; (6) False Bay.
New building has engulfed old residential areas, some occupied by Coloureds. Many families in District 6 (7) owned title to their land, but were forcibly removed in the 1960s. The area is still not rebuilt

cargoes handling about 18 million tons but Richard's Bay to the north handles about 37 million tonnes. Most ports now have modern container terminals, and some goods in containers travel by train right through to the PWV triangle.

Apartheid, a strategy for racial segregation

One of the most controversial issues relating to the Republic of South Africa is the policy of *apartheid*. Apartheid is not just about who sits on which seats, or swims from which beach. It is a vast social engineering programme aimed at confining and controlling blacks, and maintaining the privileged way of life for the white minority. It is 'geography', because it affects the pattern of settlement, that is, the spatial distribution of people in South Africa.

The term 'apartheid' is out of favour with the government. Over the years several terms have been used for areas set aside for blacks to live in: Native Reserves, Bantustans, Homelands, National States. But they all relate to a consistent policy of separate development which started with the slave policy of the Dutch East India Company in the mid-seventeenth century. It means that blacks have no right to live anywhere else. In the 1970s the first Independent Homelands were set up. They are now called Independent National States.

Discussion of apartheid is often about statistics: the numbers of blacks and whites, how much land each has and about pressure in the Homelands driving the able bodied men and women to work in white areas.

- Homelands are incapable of supporting anything like the number of people shown in the table below so black Africans are forced out.
- They are only wanted as a labour force in certain parts of white South Africa: elsewhere they are not wanted, they are *surplus people*.
- Between 1960 and 1980 3.5 million people have been moved and 1.8 million more face eviction. This is more than the total white population.

Separate development at different scales

Figure 17.13 shows how the policy of apartheid works for different groups of people at different scales.

- In the Homelands too many people put pressure on the land. Men go away for wages work, families are separated. They have no rights as citizens of the Republic of South Africa.

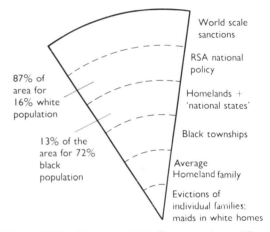

Figure 17.13 How apartheid affects people at different levels

---

Population and density (1985): who lives where

		Total people	Area sq km	Density
1	The total population of the RSA is (73% of the population are blacks)	▶ 31.3 million	1,221,000	26/km²
2	The 4 'Independent Homelands' – Ciskei, Transkei, Venda and Bophuthatswana – have	▶ 5.1 million	96,000	53/km²
3	The other 6 Homelands have	▶ 6.9 million	67,000	103/km²
4	Officially all blacks belong to a Homeland but there are	▶ 10.8 million outside Homelands		
5	This makes a total of	▶ 22.8 million blacks who are		
	supposed to live in the 10 Homelands, an area of		▶ 163,000	
6	This would give a population density of			▶ 140/km²

- On a district or neighbourhood scale it means forced removals. It means the clearing of 'black spots', that is, large residential areas like District Six or squatter areas like Crossroads, both in Cape Town.
- On the domestic level 'separateness' is the maid or gardener allowed to live over the garage of a white employer for as long as they have a job; or the squatters scattered on bits of land on white farms. If they are lucky and get 90 days work from the farmer they can stay on the farm, but get no pay. They do, however, get 'permission to plant'. It can also mean that people are cut off from their water supply.

Statistics, if they come from a reliable source show the scale of a problem and provide the setting for particular cases. The latter bring home the realities of the system and how it works for real people in the real world. First, for blacks working in white areas. Second, for blacks and others living in white areas or black spots, and faced with forced removals. Third, for families and individuals in their daily lives.

Blacks in white areas

Something has already been said about black people who work in white areas. At least 1 million women work as 'maids'. About 8 million black men and women work in white areas in some way.

These figures show the needs of the South African economic system for workers. There is a 'pull' into white sectors, and a 'push' away from Homelands and rural areas which are increasingly overcrowded and unable to provide a livelihood. Estimates suggest that the Homeland population has doubled in 20 years and that 80 per cent of people in paid work are not earning an economic wage. Where jobs are increasing in some Homelands, it is in the administrative sector: white-collar jobs, which do not help the poorest people.

Forced removals in South Africa

Many black and coloured people now live in places where the government does not now want them to be. For example:
- small 'squatter' groups on white farms
- where the land is wanted for expanding towns or suburbs. Long-established ethnic suburbs are often surrounded by white areas as towns expand. Long-standing black residents often have title to their property
- where land is wanted for development or strategic purposes, such as dams, game reserves, military developments, border areas
- where townships and informal settlements are unsightly, and give a bad impression when viewed from main road or railway. This includes satellite townships (see page 152) and spontaneous settlements like Crossroads (75,000 people) near Cape Town
- in scattered ethnic groups which the government needs to consolidate in order to make credible Homelands.

The government also wishes to force people back to the Homelands now that smaller numbers of workers are required. This is partly due to increased mechanisation in farming and industry, and also to trade recession.

> 'The worst you can do to a person, except kill him, is uproot him forcibly from his home'

All these people are surplus people, giving their name to the Surplus People Project, a report about forced removals in South Africa.
- Between 1962 and 1980 3.5 million people were moved
- Nearly 2 million people are still threatened with eviction
- These add up to more than the present white population of South Africa.

The government says that these people have often been offered prepared sites for new homes. But conditions were often less good; they had further to travel to work, or to find work, some lost jobs, all felt uprooted from their heritage.

What taking away land can mean in a rural area

During the development of the Nduma Game Reserve in northern Natal, a tall wire fence was erected through the traditional homeland of the Bangweni people. This cut local people off from their water supply in the Pongola River. Women and children were allowed to collect water by a special gate, but they could not carry enough and the results were:
- cattle were short of drinking water and could not be properly dipped so died from cattle ticks

- people lost their vegetable gardens near the river, so had to carry water to new gardens further away or buy food
- they also lost their usual thatching material
- people could not bathe so water for washing had to be carried.

What has to be remembered is that Removals often result in making people poorer (legalised impoverishment); and they lose their roots and traditional way of life.

Bophuthatswana, an example of one 'independent' National State

How many people?
- Transkei and Bophuthatswana are the largest National States. Bophuthatswana has an area of 38,000 sq km (larger than Lesotho).
- Bophuthatswana has 6 blocks of country strung out, west to east, near the border with Botswana. The southernmost, Thaba Nchu, is 450 km away between Bloemfontein and Lesotho (see Figure 17.10).
- The individual parts have very uneven levels of development.
- There are between 1.5 and 3 million Batswana in South Africa. Batswana also live in Botswana. Bophuthatswana is supposed to be the 'homeland' of the Batswana people but only 866,000 Batswana live there.
- There are also about 400,000 non-Batswana in Bophuthatswana. So this is a 'homeland' where nearly a third of the residents are non-Batswana and two-thirds of the citizens are always out of the country!

Problems of a fragmented 'independent' National State within South Africa

The central block, Mmabatho–Mafikeng lies on the railway between Gaborone and Johannesburg so it has a link to Kimberley and Cape Town and to the main south–north route to Zimbabwe.

In the colonial period Mafikeng (then Mafeking) was the administrative centre for Bechuanaland, now Botswana. When Bophuthatswana was set up, Mafikeng, a white town, remained in the Republic of South Africa. So a new administrative area had to be built at Mmabatho just to the north of Mafikeng, on the other side of the railway line. This surge of development convinced Mafikeng that it should join with Bophuthatswana. Look at the photograph, Figure 17.14.

Figure 17.14 University buildings at Mmabatho, Bophuthatswana

Thus Mmabatho–Mafikeng is a remarkable combination of historic old colonial town and completely new development with the most modern of buildings. It seethes with activity, and if this willingness to experiment can be matched by economic survival, it could become a model of racially integrated old and new.

Can Bophuthatswana survive? Some questions to ask:
- Can Bophuthatswana overcome the problems of being such a fragmented state? The key problems may not be in the Mmabatho–Mafikeng block, but in the other parts of Bophuthatswana.
- Could Bophuthatswana ever support all its 'citizens' if they should return to the ethnic homeland?
- About 40 per cent of Bophuthatswana's income still comes from the government of the Republic of South Africa. Much of the rest comes from mining (particularly platinum), game parks and leisure centres like Sun City.
- The new administration is working hard to develop Bophuthatswana as an independent state. But if the governments of other countries do not recognise it, are its citizens in 'no-man's-land'?

Land, food, and people: the agricultural use of the land in South Africa

The simplest division of landscapes to remember for South Africa is:
- the vast, high interior plateaus
- the coastal lands of varied width and character
- the Great Escarpment, separating these two and reaching nearly 3,500 m in the highest parts of the Drakensberg (see Figure 17.15).

These 3 divisions are modified by local features (for example the Karoos in the Cape) and also by climate, to produce 8 or 9 distinctive areas. The varied conditions encourage contrasts in land use and a great variety of commercial crops. Terrain and soils change but the main limiting factor is the rainfall or the availability of water for irrigation.

The area of cropped land is much smaller than the vast areas of dry grazing land and scrub. But travelling along the railway from Johannesburg to Cape Town in April, there are huge cattle ranches, maize for mile after mile, and sometimes sunflowers, sorghum or fodder crops. In places grasslands stretch to the horizon and seem under used. Everywhere it is flat, flat, flat, with only an occasional ridge in the distance. These were the vast 'empty' grasslands or veld across which the *trekboers* drove their wagons. It is perfect railway building country. Sometimes there are man-made hills: the spoil tips of Klerksdorp and other mining centres. Every town-stop along the railway has its grain silos.

How important is agriculture? No agricultural product appears in the table of exports on page 145. But agricultural *production* is a different story. The 'gross value' league table is:
- first, livestock products: meat, eggs and poultry, dairy products (total R3,500 million);
- second, grains: maize R700 million, wheat R650 million, hay R500 million;
- third, horticultural products (R600 million) including a great variety of fruit and vegetables. In addition there is greatly increased value when grapes are processed as wine;
- fourth, cane sugar R450 million.

Livestock products, a great money earner

This is the most extensive farming activity in South Africa. Production figures are only given for white areas, not for traditional stock raising.
- There are some concentrations of dairy farms near the urban, industrial and mining centres. These include the PWV metropolitan area, the midlands of Natal, east and west Cape Province and east and north Orange Free State. There are smaller concentrations of beef cattle.
- The Karoos are the main sheep farming areas, with some others near markets in dry areas. About 66 per cent of the sheep are Merinos producing very good wool. *Afrinos* are meat and wool producers adapted to dry conditions.

Commercial grain farming

Nearly half of the cultivated land of South Africa is planted to maize. Maize is an outstanding crop on the drier plateau, particularly in the area known as the maize quadrangle. Figure 17.16 shows a small part of a farm in the Orange Free

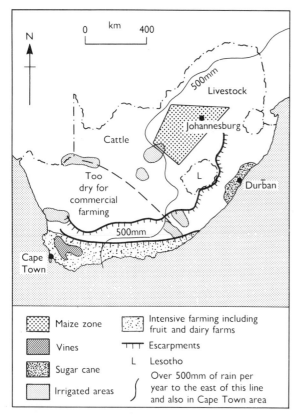

Figure 17.15 Landscape, agricultural regions, and commercial crops in South Africa

The Republic of South Africa 157

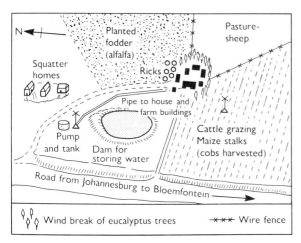

Figure 17.16 A Highveld plateau farm in the Orange Free State. Height: 1500m. Rainfall: about 600mm per year. Crops: maize, oats. Stock: cattle and sheep.

State near the southern limit of the maize-growing area. The original farm units were a huge 2,500 ha but most have been subdivided. Crops are diversified to include oil seeds, including sunflowers.

As the annual rainfall decreases on the plateau to the west and south-west, crop farming gives way to grazing, though crops are still grown to feed the stock. Figure 17.15 shows where commercial crops are grown in South Africa. Irrigation is being widely developed for speciality crops. Commercial crops are grown in the south-west Cape within 150 km of Cape Town and the southern coast valleys of the Cape Ranges, mainly in the winter rain area. There are Outspan citrus farms in the eastern Cape as well as others in northern Transvaal. Some of the most important commercial crop land is found on the east coast in a 400 km strip in the vicinity of Durban (see page 158).

Lourensford: fruit farming in winter rain areas

Pears may not be mentioned in trade figures but the Lourensford study is of value because it spells out:
- the enormous complexity of agribusiness in the modern world;
- some details of farming in a 'mediterranean' climate area where there is no rainfall during the summer growing season from November to April.

The use of the land at Lourensford

	Hectares
Total area	3,860
Pine forest	1,400
Pears	237
Apples	170
Plums	22
Lemons	14

There are 2 main problems: water and heat. Winter rain must be stored in dams and altitude and aspect used to reduce temperatures.
Now use Figure 17.17 to find:
1 What has been planted on the steepest slopes to reduce rockfalls, regulate water flow, and reduce wind force.
2 The sequence of fruit orchards down the slope. Suggest reasons for their position.
3 Irrigation furrows (open cement-lined channels) which follow the contour and water the orchards. Underground asbestos pipes from

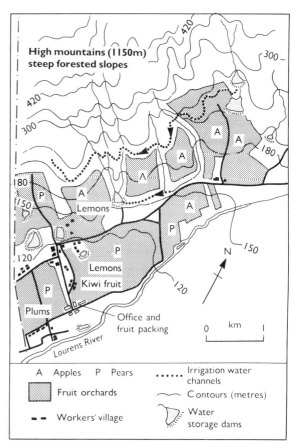

Figure 17.17 Lourensford estate, Cape Province

158 South Africa

the dams have valve outlets in the orchards.
4 The cooler southerly aspect. This reduces temperatures but it is still hot for temperate fruit. Britain grows apples and pears. Compare the two latitudes.

Lourensford today

Those of you who have an earlier edition of this book will be able to check the changes that have taken place. Peaches used to be the main crop but demand has changed.

About 2,000 people live and work at Lourensford for much of the year, some of it on seasonal contract. Coloured workers total over 500 men and women, and there are about 400 black workers. This means that there is a central village with a store, soccer and rugby pitches, primary school, day nursery and a clinic, as well as two smaller villages with housing for workers and their families. Older children cycle to school in Somerset West about 8 km away.

Lourensford is both a family and a company estate. In less than 50 years it has expanded and taken care of most of its infrastructure needs. It has a saw mill, a fruit tree nursery, the apiary to ensure a supply of bees for pollination at the right time, a dam-building team that also maintains water courses, a research unit, and of course organisation and marketing teams, summed up as 'management'.

Cane sugar production along the Natal coast near Durban

The cane sugar industry of the Natal coastlands is significant for today's Republic of South Africa. The need for a large labour force resulted in the recruitment of indentured workers, mainly from India. Their descendants form the main basis of the Asian community in South Africa today. The system of growing cane sugar in Natal and on other traditional estates is very different from the one already studied in Kenya (Mumias, pages 52–54).

Check the conditions needed for growing sugar cane in the fact box on page 54 and the notes and diagram in Chapter 1 (page 10). Study the two following sections
• Figure 17.18 showing the position of the Tongaat sugar estates and text on sugar cane growing and processing.
• Practical work box 8, including Figure 17.19 which is an aerial view of part of the Natal coast area, growing sugar, and Figure 17.20, a line drawing.

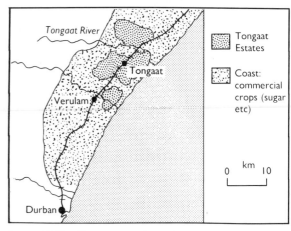

Figure 17.18 The location of sugar cane estates at Tongaat, Natal

1 Compare the different systems
2 Debate the reasons and the advantages and disadvantages.

The Tongaat sugar estates

Tongaat is a small sugar town 40 km north of Durban on the coast road and railway.

Figure 17.15 and the map of the Tongaat area, Figure 17.18, show that the main growing districts stretch from the sea inland for 15 km, an important factor in reducing the likelihood of frost. Rainfall is only about 1,100 mm a year along much of the Natal coast. Sprinkler or spray irrigation is used.

Huge tractor-trailers collect cane at the 'cutting face' and take it to farm roads where heavy motor trucks move it to the mill.

Sugar cane growing and sugar production is a highly mechanised, large-scale business. Although Tongaat has diversified into other business and agricultural interests, cane sugar production is still largely a monoculture.

South Africa earns about $100 million a year in exporting sugar, much of it to Japan. There is a huge bulk sugar loading terminal at Durban which can load a 10,000 ton vessel in less than 24 hours. But like other sugar producers South Africa has to compete in a world with a depressed market for sugar, hence the importance of the increasing home market.

Secondary industries based on sugar give work to over 100,000 people. Sugar is also used in the manufacture of shoe polishes, adhesives,

The Republic of South Africa 159

Practical work box 8: a Natal sugar estate: interpreting air photographs

Figure 17.19 An air view of a sugar cane estate north of Durban

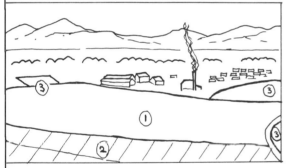

Figure 17.20 A landscape sketch based on Figure 17.19

Introduction
- Aerial photographs are special because they give a much better view of the landscape.
- Figure 17.19 shows an air view looking across sugar estates in Natal. It shows an oblique view looking away into the distance. Sometimes an aerial photograph looks straight down onto the road pattern of a town or village. It is then called a vertical photograph.
- If you use speed drawing, a few lines will give a landscape sketch which can be annotated (see Figure 17.20).
- The word interpret means to study something carefully so that you understand it and are able to explain it.

Interpreting Figure 17.19
Find the following:
- Foreground: sugar-cane fields
- Centre: sugar-cane processing mill (note smoke from power house)
- Centre right: town, neighbourhood
- Middle distance left: another estate?
- Low foothills: small farms? African farming? Is this land higher than the cane fields?
- Horizon: higher hills or mountains. How far away?

Headings like these are useful for any photograph study. They suggest simple groups or categories of land use. The question marks indicate places where features are not easily identified.

Now look more closely at Figure 17.20
1 There are two patterns in the cane fields. There are some very large fields (1) but the nearer ones are subdivided (2). What are the reasons for this?
2 What may be happening where (3) is marked? (3 places)

photographic materials, pharmacy products and explosives. Molasses can be used as a fertiliser and feed for stock. Bagasse, a waste product, is used as fuel in the sugar factories and paper can be made from residues.

Water resources in South Africa

Despite the existence of great rivers such as the Limpopo, Vaal and Orange, South Africa is not well endowed with water (see Figure 17.21).
- Almost half the country is arid or semi-arid. The western third includes desert and has annual totals of less than 100 mm to 200 mm.
- The eastern part of this area has more rain, up to 400 mm, but the climate is unreliable.
- Most of the eastern third of the country has a rainfall of 500 mm or more.
- Only the east coast (1,000 km on either side of Durban) has over 1,000 mm of rain a year.
- The southernmost coastlands have between 600 and 800 mm average annual rain.
- Everywhere there is a dry season that may be as long as 6–9 months.
- Everywhere except the southwest Cape the rainy season is also the hot season.
- So evaporation (evapotranspiration) exceeds rainfall. It reduces the value of the rain for rain-fed crops and irrigation, by a loss of up to 40 per cent from the surface of rivers, lakes and reservoirs.
- Even in the west Cape the rainfall is seasonal (April to September) producing a 'Mediterranean' climate.

Limited water supplies not only reduce agricultural production. Industry uses immense amounts of water, and so do people in towns. Johannesburg has used so much underground water from the local rocks that polluted water is getting back in.

South Africa's spending on water engineering is vast. Nearly 400,000 ha of land is irrigated, a four-fold increase in 25 years. There are 2,000 km of irrigation canals. Some of them transfer water from one drainage basin to another through tunnels. The following are some of the chief schemes:

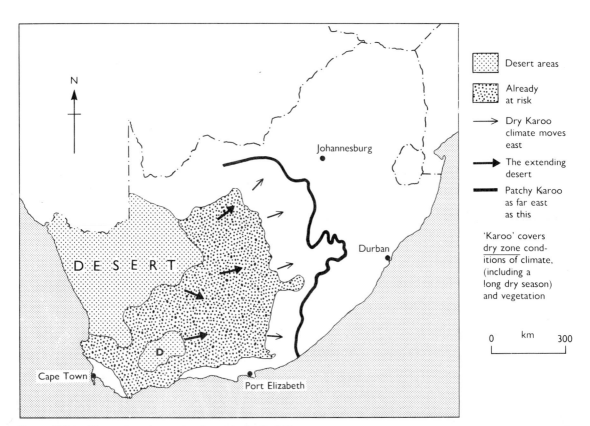

Figure 17.21 The eastward creeping desert in South Africa

- the Orange River Projects, started in the 1960s
- the Hendrik Verwoerd Dam system
- the Orange-Sundays-Fish rivers schemes

But there are 2 important new proposals. It looks as though the Highlands Water Scheme, will go ahead after being talked about for 20 years; the other may never happen, but if it does, it could transform some of the driest parts of southern Africa.

1 *Lesotho's white gold': the Highlands Water Scheme*

Lesotho includes some of the highest parts of the Drakensberg Mountains. Its heavy rainfall is a great untapped asset. Water from Lesotho will be sold to the Republic of South Africa and could double the flow of water into the River Vaal. Lesotho will gain revenue and also a hydroelectric scheme. This huge water engineering scheme will involve 5 dams and a tunnel nearly 100 km long. It will be completed by stages, stretching well into the twenty-first century.

2 *'Zambezi water for the reef'*

Some writers say that the major metropolitan area of South Africa, the PWV triangle, is desperate for water. A German water engineer has proposed bringing water from the Zambezi, before it goes over Victoria Falls. A 1,000 km-long canal or pipe would tap the river near its confluence with the Chobe river and bring the water via southern Botswana. It would give more water than the Lesotho Highlands scheme, but the mind boggles at the cost and the engineering and environmental problems. It would be so costly that water could only be used for industry and domestic consumption. It would not increase the irrigated area.

The future

This chapter began by saying that apartheid was the dominant feature of the geography of South Africa. But the apartheid of today is very different from the apartheid of the 1950s–70s. There are already fairly substantial changes and some people say that 'laws are rotting away at the edges'. Some of the machinery of apartheid is being dismantled. But the change, and the resistance to change, varies from area to area and group to group. Most of the statements that follow need to be qualified by the comment: 'This happens in some places but not in others.'

So what is different? Is it real change or just improving appearances?
- The reservation of certain jobs for whites only has been dropped and more blacks are getting training for skilled work.
- There are now more and racially mixed trade unions.
- The major English-speaking universities are now 'open' and free to admit whoever they wish. Many private schools are now multiracial, for example, the Roman Catholic church schools in large cities.
- Mixed marriages are now legal.
- There are more independent black businesses, and the business districts in some towns now offer trading for all races.
- Blacks can now hold 99-year leases on their homes in some areas.
- There has been relaxation of some petty regulations such as separate bus seats, beaches or toilets; and there is mixed sport.
- Limited voting rights for a central national parliament have been given to coloureds and Indians but not to blacks.
- The Pass Laws have been abolished.
- There is a strong white group opposed to apartheid.

What future path?

The situation in South Africa cannot stay as it is: no-one is satisfied with it. It is too easy to say all black people want this, or all white people think like that. They do not.

Almost any proposal faces problems:
- Blacks are divided by tribe or nation but also there are large numbers of urban blacks who have more in common with each other than with their tribe. There are few compact tribal 'living spaces'. In some areas ethnic groups overlap and claim the same land. Violence results from this as well as attacks on blacks who work with whites and are said to collaborate.
- Some black households have easier access to jobs, especially in metropolitan areas. For example, some Sowetan families earn more money, improve their homes and send their children to private schools. They become 'insiders'. Many blacks are better off financially than people in other black ruled African counties. This creates a gap between black 'haves' and the 'have nots'.
- People living in rural areas, especially in overcrowded homelands, are often the 'outsiders'. They really are on 'subsistence'. Their men and young people may be away trying to earn money. They are less likely to have good schools, housing,

health services or have cash and manpower to improve their farms. They have very little to lose.
- There are big gaps in training and experience between whites and blacks. All groups need access to business management schemes, and advisory, and marketing schemes for farmers. So far they benefit the whites and increase the gaps.
- People have different political ideas about what is best for a country and how this can be organised and achieved. It takes a great deal of sophistication and wisdom to prevent these political differences from developing into violence (see any history book).

Think about these points
- South Africans are such an ethnic mix, and have been for hundreds of years despite racial classification, that only a society organised on multiracial lines will work.
- All South Africans should be able to move freely about their country.
- All South Africans should share in government and have common citizenship.
- The sharing of resources and privileges must be seen to be reasonably fair. But in an unequal world one cannot hope for equality immediately.
- There should be solid guarantees for the rights of minorities, including the white tribes. There are nearly 3 million coloureds and nearly 1 million Asians to be considered.
- People will only invest money in businesses if there is political stability. Continued economic growth could benefit all groups.
- All South Africans should be able to seek jobs in any market and get suitable training. Job mobility is essential for any economy to work well.
- If integration is too difficult for a simple majority government solution, could a federal system, or a cantonal system as in Switzerland, be made to work and be more acceptable (see Figure 17.22)?

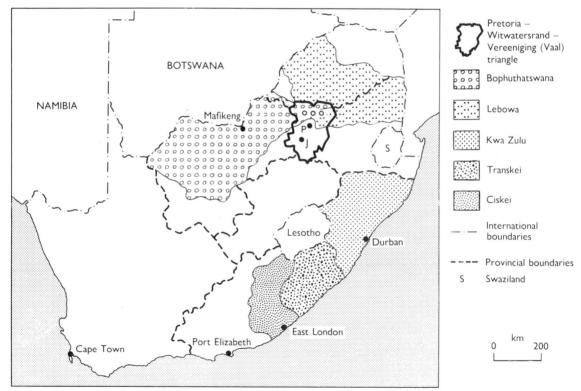

Figure 17.22 Regions for development planning.
- This map combines old and new situations.
- It shows the areas claimed for ethnic homelands about 20 years ago. In the 1960s and 70s, Black Homelands said they would aim at territorial federation in the long term
- Boundaries are also shown for National Planning and Development Regions. These boundaries are different from the present provincial boundaries for Cape Province, Natal, Orange Free State and Transvaal. Use an atlas to check the differences.
- Could this map form the basis of a federal system like the USA or a cantonal system as in Switzerland?

The Republic of South Africa

Lesotho

Capital: Maseru, 45,000 (1976)

Exports $24 million
Imports $426 million
Visible trade balance $402 million deficit (1984)

Export commodities	%	Export partners	%
Diamonds	56	South Africa	34
Mohair	11	EEC	11
Wool	9	Rest of Europe	55

- A land-locked country completely surrounded by the RSA.
- A spectacular mountainous country, above 1,000 m, occupying the highest part of the Drakensburg escarpment.
- Only 10 per cent of land is cultivable and 70 per cent of people are stock raisers.
- 23 per cent of the labour force work in the RSA mines and send money back.
- Diamonds are the only valuable export.
- Water resources may be developed for sale to RSA.

Swaziland

Capital: Mbabane, 39,000 (1982)

Exports $272 million
Imports $351 million
Visible trade balance $79 million deficit (1984)

Export commodities	%	Export partners	%
Sugar		South Africa	32
Wood Pulp	38	UK	22
Fertilisers	14		
Canned & fresh fruit	13		
	12		
Asbestos	6		
Timber	3		

- A land-locked country lying between RSA and Mozambique.
- High mountains in the west with high rainfall, more moderate rainfall in the east.
- Soils and rainfall are suitable for a variety of crops: sugar, cotton, fruit and rice.
- Asbestos and iron ore are mined.
- Vast areas of forests have been planted and timber and pulp products are important exports.
- A member of the South African Customs Union but has a rail link to the sea at Maputo (Mozambique).

PART 3 The Continental View

Chapter 18 African resources and development

> **Key words**
>
> Unity, common needs, potential, integrated and sustainable development, wealth

It is very hard to get a true picture of a whole continent. General statements can nearly always be both *right* and *wrong*. They are almost certainly true of one part but not another. If this is so, why should we try to get a continental view?

One reason for looking at Africa as a whole is that the countries of Africa have already set about establishing a more united Africa through the OAU (Organisation of African Unity). This does not always work smoothly. But it does give a meeting place for getting together to sort out problems.

This book studies the whole continent, not just Africa south of the Sahara. Many important developments are influenced by situations in the northern part of the continent. Islam overflows into West Africa and far down the east coast. Water regulation projects on the upper Nile influence what happens in Egypt, and an open Suez canal makes a difference to ports in eastern and southern Africa.

Africa shows promise of being possibly the first continent to be 'planned', if not as a whole, at least from a regional viewpoint. There are some situations common to many of the countries of Africa:

- Some areas have features in common, such as their climate or agriculture and produce similar exports which compete for markets.
- Groups of countries need to tackle common problems such as refugees, locust swarms, or transport for land-locked states.
- Research is more economical if it is tackled from this wider view instead of each country trying to fund its own. In any case, African countries just have not got enough money to spread around wastefully.

Two money problems that many African countries share are:

1 Foreign exchange is often earned from export crops. The value of these goes up and down according to world markets.

2 Much of their foreign earnings is used up in paying interest on loans. Many countries have borrowed heavily to fund development projects, build schools, etc. They could not know that there would be a world recession, increased prices for the oil they have to import, and terrible drought.

Every country wants to know what its future prospects are and how to *realise its potential* that is, how to make the most of its resources. Over the last 30 years people's ideas have changed.

- In the first days after independence most countries' plans were concerned with increasing agricultural production, and land reorganisation.
- Then it was the turn of the large project with paper plans spread over five years or more. Some of these have still not been completed.
- Now countries try to decide targets and have a more or less continuous planning strategy. The focus of this is often a three year 'rolling plan' which is adjusted each year according to what has been achieved so far, and what is now seen as priority need.
- There is also a move towards provision of essential services for the whole population: schools, clinics and rural water supplies.
- People are more aware that everything connects: population with food supply, soil erosion with over-grazing and farming steep slopes. There is a move towards *integrated* development.
- It is often better to make small but lasting improvements which are *sustainable* forms of development.

There is also a realisation that regions cannot be 'equal' because they are differently endowed and develop at different rates. The map on page 204 shows Areas of Opportunity in Africa. We are now in a much better position to know why each of the places marked offers opportunities to people. These places are also the 'eonomic islands' of the continent, the places where resources are being developed now.

Practical work box 9: how to draw a simple pie chart (graph)

One of the most difficult things for a government to decide is how to *divide* up the money it has to spend on modernisation and development.

1 Suppose it decides to use:
- *one-quarter* on upgrading of rural areas (service centres, new schools, clinics)
- *one-quarter* on housing improvement in informal urban settlements
- *one-quarter* on industrial development (a pulp and paper mill; the small industries sector)
- *one-sixth* on infrastructure (roads; double tracking single track railways; rural electricity)
- *the remaining twelfth* on contingencies (things that turn up unexpectedly)

2 If the government wants to show this visually to the general public it can do so through a pie chart. This shows the proportion (*slice of the cake*) given to each sector.

3 *Method (see Figure 18.1)*
- Draw a circle (use a coin as a guide).
- Divide it into quarters.
- Then divide it up into 12 sections like a clock face (not a digital clock!)
- Use the information given above to work out the proportions in twelfths ($\frac{1}{4} = \frac{3}{12}$, $\frac{1}{6} = \frac{2}{12}$).
- The fifth diagram in Figure 18.1 is another way of showing the slices of cake given to the different spending programmes.

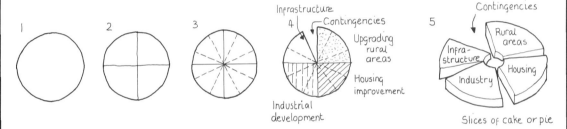

Figure 18.1 How to draw a simple pie chart

Economic development and wealth

Economists measure the 'wealth' of a country by adding up the total value of all goods and services that are produced and sold, both within the country and abroad. This is called the gross national product (GNP). Most African countries have a small population and a small GNP. The 5 countries with the largest GNP account for nearly 70 per cent of the total for the whole continent (see table below). The remaining 32 per cent is shared by over 40 countries.

The GNP may be shared between many people, as in Nigeria (92 million) or relatively few (Libya 3.6 million). When the total GNP is divided by the number of people, it gives the average product per person (that is, *per capita*). This can be used to compare the wealth of one country with another.

The traditional way of life has much to offer, but economists would not be able to measure much of it. 'Subsistence' or domestic agriculture can feed a family but does not result in cash that can be counted. A homestead with clay brick walls and a thatched roof, built with the labour of an extended family, does not need a building firm whose income appears in the accounts of the tax inspectors and national economists. Thus gross national product covers only part of the many transactions that take place in a country.

People living in a country with a high *per capita* income may not necessarily be leading better lives than those elsewhere. The difference

Percentage of the total GNP of Africa produced by:	
South Africa	20.5
Nigeria	18.9
Algeria	13.1
Egypt	8.5
Libya	6.7

is that goods are being produced that can be sold for cash which can in turn be used to purchase other things. This is true at the national as well as at the domestic level.

> 'They give us 30 tons of sorghum, 10 tons of rice, but we eat that up and it's gone. We have been asking for years for help to build dams so that we can feed ourselves.'

The amount of benefit to ordinary people depends on how the money is spent and how the wealth is shared. Some countries are spending more and more on weapons and military equipment. Many countries have their 'Wabenzi' (from Mercedes-Benz cars), a small, rich elite group. Sophisticated and comfortable living for a few can quickly use up money that could bring better health and a more balanced diet for many.

Although GNP *per capita* is not necessarily a good measure of human well-being, it is instructive to compare those countries which have a high *per capita* income with those which do not, and to see where this wealth comes from.

High and low income countries

The wealth of most African countries comes from producing commodities for export. Information on exports appears in the trade summaries in Part 2 of this book.

Figure 18.2 lists the 10 countries in Africa that have the highest (group A) and lowest (group B) *per capita* incomes. It shows their trading position (size of surplus or deficit), and the two most important export commodities. It is based on GNP data for 1983 and export data for 1982–84. Several interesting points emerge from this table.

1 Minerals
- Minerals, especially oil, play a large part in producing the wealth of the countries with the high *per capita* incomes (group A).
- The relative positions of the top 4 countries have remained unchanged since 1975, and 3 of them get most of their wealth from oil.
- Cameroon, Namibia, Botswana, and Swaziland have entered the group of rich countries since 1975, with minerals providing the wealth in 3 of these countries.

2 Agricultural products
- Group B, the poorer countries, are mostly dependent on agricultural production. They may have mineral deposits, but they have not been developed.
- A depressingly large number of African countries have very low incomes.
- Because agricultural production is unreliable (weather conditions, price drops) there are frequent changes in the membership of group B.

3 Population
- Several of the high *per capita* income countries have very small populations. The benefit of a valuable resource is being shared among relatively few people.
- Both Nigeria and Egypt have a very large GNP based on oil production, but do *not* appear in this table because the wealth is shared among a large population.

4 Location
- The richer countries are coastal, or situated in the southern part of the continent. Many oil deposits are found in the shallow offshore waters, and southern Africa has very valuable minerals, especially gold and diamonds in the underlying Basement rocks.
- Several of the poorer countries such as Burkina Faso, Mali, and Chad are in savanna areas where the water supply limits the possibilities for agriculture.

5 Market prices and balance of trade
- Niger was one of the poorer countries in 1975. The development of uranium deposits provided an income which lifted Niger out of the bottom group of countries by 1981. Since then, there has been a dramatic decline in the world market price for uranium, and Niger has reappeared in the lower part of the table.
- Similarly Zambia has dropped out of the top 10 group because of the much lower world price for copper.
- Some of the high income countries have trade deficits. It is possible to be rich, but still to overspend!

Figure 18.2 High and low income countries in Africa

Country	Income per head $ (1983)	Visible trade balance	No. 1 export	% of country's exports	No. 2 export	% of country's exports
Group A: High Income Countries						
Libya	$7,500	$1,977 m surplus	Petroleum	100	—	
Gabon	$4,250	$1,274 m surplus	Petroleum	81	Timber	8
South Africa	$2,340	$1,079 m surplus	Gold	47	Base metals	10
Algeria	$2,400	$3,226 m surplus	Petroleum	99	—	
Namibia	$1,760	$4 m deficit	Diamonds/Uranium	(details not available)		
Tunisia	$1,290	$1,116 m deficit	Petroleum	45	Clothing	12
Congo	$1,230	$416 m surplus	Petroleum	90	Timber	5
Botswana	$920	$119 m surplus	Diamonds	72	Copper/nickel	8
Swaziland	$890	$79 m deficit	Sugar	38	Wood pulp	14
Cameroon	$800	$84 m surplus	Petroleum	42	Cocoa	12
Group B: Low Income Countries						
Tanzania	$240	$350 m deficit	Coffee	39	Diamonds	23
Niger	$240	$302 m deficit	Uranium	79	Live animals	12
Uganda	$220	$49 m deficit	Coffee	97	Cotton	1
Malawi	$210	$28 m surplus	Tobacco	52	Tea	27
Guinea Bissau	$180	$37 m deficit	Shellfish	30	Groundnuts	24
Burkina Faso	$180	$234 m deficit	Cotton	42	Karite nuts	17
Zaire	$160	$605 m surplus	Copper	42	Manufactures	16
Mali	$150	$78 m deficit	Cotton	54	Live animals	15
Ethiopia	$140	$432 m deficit	Coffee	61	Hides	10
Chad	$80	$21 m deficit	Cotton	91	Live animals	1

Trade data are mainly for 1982–84; South Africa trade data is for SACU.

Resources and development

The focus of Part 3 is on resources and alternative ways of using them. The next 4 chapters concentrate on:
- people as a resource
- the land and the rural way of life
- urban life
- industrial development and infrastructure, that is, the support system that makes economic development possible

The last chapter refers again to the common situations that affect the whole continent, and the planning and political aims that can improve them and help to move Africa out of the difficulties of the late twentieth century into a hopeful twenty-first century.

Chapter 19 People as a resource

Key words

Population pyramids, balance of birth and death rates, disease as a hindrance, wasted lives, women's contribution

- However rich a country is in natural resources they are useless until *people* work on them. Without people and their energy and ideas, development cannot take place.
- When governments make plans they must know how many people they are planning for, where they live and other relevant facts. For example, if the school-age population is going to double in 10 years, schools must be built in time.

Counting people

A count of a country's people, a census, is usually taken every 10 years. It is very costly and takes teams of skilled *enumerators*. When the count has been made, the census forms have to be processed. Even when sophisticated computers are available the analysis is long and difficult. Only when the results are known can governments begin to see changes and plan accordingly.

One of the best ways to understand the balance in a country's population is to see it in a *population pyramid*. This shows graphically the numbers of people in each 5-year age group, and their sex. A population pyramid has already been used to contrast the number of men and women in each age group in a communal area of Zimbabwe and in Harare (see page 138). It shows how men going away from home to work in a town changes the *balance* of the sexes in both places.

Study the three pyramids shown in Figure 19.1 and see what you can learn.

Figure 19.1: explanation
1. Each *age band* of population accounts for five years. For example, the age band 15–19 includes people who are 15, 16, 17, 18 and 19 years old.
2. The lowest 3 bands (called 'cohorts') are shaded to show the young people. Many of them are in school.
3. The bands of the elderly are also shaded.
4. In between is the main working population.
5. The base line shows percentages, that is, the proportion of the total population in each age band.

Figure 19.1: interpretation
1. A wide base means many babies born and young children surviving (A).
2. A wide top means many older people surviving and perhaps living longer (C).
3. A pointed pyramid means a high death rate, that is few people live to be very old (B).
4. If the middle of the pyramid is wide, more children are surviving to become teenagers and adults (C).
5. A very narrow base compare with the rest means that the births are less than in previous years (C).

Practical work box 10 on page 170 explains how

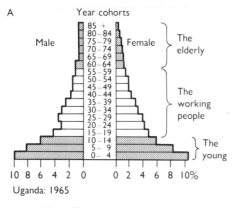
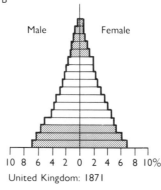
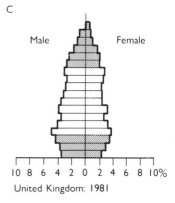

Figure 19.1 Three population pyramids

to draw population pyramids. Figure 19.2 shows simplified pyramid shapes.

Too many people

Population growth is slowing in many parts of the world: but not in Africa. Some African countries have the highest growth rates in the world (3.5 per cent).

Two things together cause a population surge: a high birth rate at the same time as a lower death rate. This occurs when more children are born, fewer babies die, and older people live longer.

Figure 19.3 shows very simply how the population of a country increases if death rates fall at a time when births are about the same.

1. At stage 1, births and deaths balance so total population remains the same.
2. At stage 2, births continue at the same level but the death rate falls, so the population rises.
3. At stage 3, the birth rate falls but births still outstrip deaths. Population continues to rise.
4. At stage 4, the birth rate and death rates level out so the population is high but more stable.
5. Most African populations are at stage 2, with rapidly increasing populations.

If the number of people in a country continues to rise, production of everything must also rise.

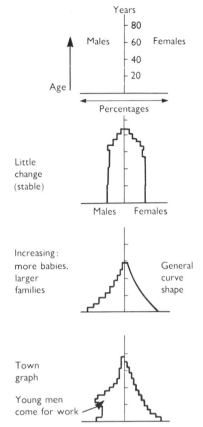

Figure 19.2 Population pyramid shapes

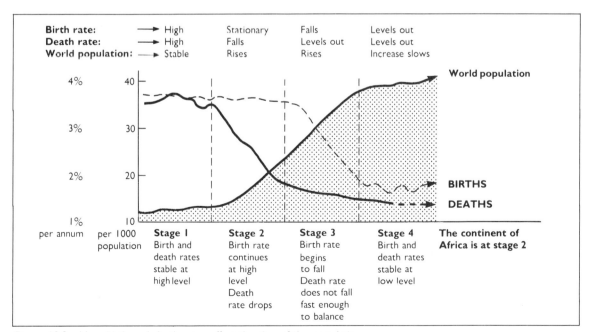

Figure 19.3 How birth and death rates affect the size of the population

> ## Practical work box 10:
> ### how to draw a population pyramid
>
> You can work out a population pyramid for yourself by using the population figures for Botswana shown here. This is a very young population so the shape of your pyramid should look like Figure 19.1A.
>
> 1 It is easiest to put the age-groups in the centre, see Figure 19.1. Many pyramids do not number each age band (called a cohort); they leave you to count up, which is not easy.
> 2 Put in a pencil base line. Measure along each side of the base. Choose an easy horizontal scale and mark it at intervals from 0 per cent to 10 per cent on each side. It is easier to be accurate if you use graph paper.
> 3 It is usual to put males on the left, females on the right-hand side.
> 4 Choose an easy vertical scale. For five-year cohorts, you usually need about 18 divisions.
> 5 Read off the percentage for the first age band (0–4 years) from the table, first for males. Mark it on the left of the base line. Do the same for females – mark it on the right of the base line. N.B. the first (top) line you read in the table is the *bottom* line of the pyramid.
>
> | Botswana
Age bands
Years | Males | Age/sex structure (1981)
% of total population
Females |
> |---|---|---|
> | 0–4 | 9.7 | 9.9 |
> | 5–9 | 8.1 | 8.1 |
> | 10–14 | 6.3 | 6.4 |
> | 15–19 | 4.7 | 5.3 |
> | 20–24 | 3.6 | 4.8 |
> | 25–29 | 2.9 | 3.6 |
> | 30–34 | 2.2 | 2.7 |
> | 35–39 | 1.7 | 2.0 |
> | 40–44 | 1.5 | 1.9 |
> | 45–49 | 1.4 | 1.8 |
> | 50–54 | 1.2 | 1.5 |
> | 55–59 | 1.2 | 1.4 |
> | 60–64 | 1.0 | 1.0 |
> | 65 and over | 1.9 | 2.3 |

If governments are trying to provide more roads, schools, homes, clinics or food and safe water, they cannot satisfy demands on present budgets, and the income 'cake' has to be divided between more and more people.

If there are more students leaving school will there be jobs for them? A totally unemployed man or woman contributes little to the product of a country. They become dependents. This is why it is often better to have labour-intensive projects rather than capital-intensive ones. By this we mean projects using a lot of manpower rather than a lot of money.

Disease as a hindrance to happiness, progress, and development in Africa

People are a continent's most valuable resource if they are well and happy and can contribute to its development. But there are several world misery-makers. War is perhaps the worst because it is entirely men's own fault. Then there are hunger, and disease. Tropical countries are at risk from a great variety of diseases related to tropical environments.

Illness brings misery to individual people. But sick people who cannot work are a liability. Most of us remember the misery of being ill and the relief of getting better. But there are some people who are ill most of the time. They take their illnesses as normal because they have never been well enough to know the difference.

The geography of disease and pestilence

Some people wonder if health, or ill-health, is a suitable subject for a geography book. But if people are a resource then anything in their physical or man-made environment which stops them working is important. Ill-health, or crops destroyed by lcousts or other pests, wastes a country's wealth.

Jacques May says 'we only get to know the travel habits of diseases when we see them plotted on maps.' This is important work for geographers. Look at the fact box, page 172. Figure 19.5A shows where locust swarms are breeding and the areas they are likely to affect. Figure 19.5B shows the areas affected by malaria and yellow fever, carried by mosquitoes. Figure 19.5C shows the areas affected by sleeping sickness and the cattle disease, nagana, carried by tsetse flies.

Unfortunately, development projects sometimes increase disease hazards. Bilharzia (schistosomiasis) is spreading as new areas are irrigated because the host snail lives on the weeds in the irrigation ditches. Ninety per cent of the people of Egypt are infected, and 60 per cent of people in some other parts of Africa. See Figure 19.5D for information about bilharzia.

Clearing bilharzia from all streams, rivers, lakes and irrigation channels and curing all the people who have bilharzia are tasks too great for a country on its own. International action is needed. It is worth knowing about the international agencies of the United Nations.

WHO	World Health Organisation
UNICEF	United Nations International Children's Emergency Fund
FAO	Food and Agriculture Organisation
UNESCO	United Nations Educational, Scientific and Cultural Organisation

Their teams can carry knowledge gained from research to people in distant villages and towns.

Sometimes local people can help in the battle by keeping their own records. River blindness caused by the small black simulium fly is common in northern Ghana and affects 20 million people in Africa. There, records were kept by community groups and secondary schools as part of their geographical fieldwork. They showed that the breeding period was the start of the rains, not the low-water season when there are many stagnant pools.

There are many other diseases such as yaws, trachoma and now AIDS (acquired immune deficiency syndrome). The AIDS virus is affecting millions of people in Africa depriving their countries of their work and adding to the burden on healthy people and the health services. AIDS may soon begin to have an impact on the balance between death rates and birth rates.

Poor diet (malnutrition) causes protein deficiency diseases such as kwashiorkor.

Water acts as a breeding ground for bacteria causing diarrhoea, a major cause of death in babies and young children. It has been said that 'there is no safe water in Africa'.

Pests and diseases also attack plants:
- Much of the world's food is lost because it is eaten by insects, grubs, rats or birds.
- Most people know about locusts (see fact box). This is an example of how a pest thought to be under control breaks out again.
- 15–20 per cent of Egypt's cotton crop can be lost through pests.
- Spraying can kill other insects which would normally keep aphids and spiders under control as well as many harmless creatures.

Check the facts in the box on page 172.

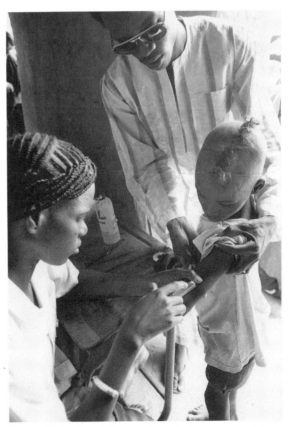

Figure 19.4 Preventive medicine in Mali. Health visitors use a schoolroom to vaccinate children

Facts: Pests and the Diseases they carry

Locusts
- Large grasshopper swarms can be so large that they can eat in one day crops that would feed half a million people for a year.
- Now 5 different species are multiplying at the same time (see Figure 19.5A).
- They need rain and protective vegetation to breed. Widespread rains after a period of drought are ideal.
- A female lays 70 eggs into the ground. 98 per cent may hatch. After 14 days they are 'hoppers'.
- Swarms of locusts were less severe from the 1960s until a sudden surge in the 1980s.
- Immense international efforts are being made to destroy breeding grounds. Cooperation between countries is essential.

Mosquitoes
- Anopheles mosquitoes carry malaria and yellow fever by biting an infected person, sucking their blood and then infecting the next person they bite. Figure 19.5B shows the areas affected.
- They breed in still water: even an old tin can will do.
- Malaria kills many children and makes adults weak and unable to work.
- Malaria seemed to be under control for 20 years but now there is a new epidemic. New breeds of mosquitoes resist the old drugs and insecticides.

Tsetse fly
- Tsetse flies bite both humans and cattle: they carry a parasite called a trypanosome. This kills domestic animals and makes people ill by damaging the red cells in the blood.
- Tsetses live in the bush, and game animals act as a 'reservoir' of infection. See Figure 19.5C for areas affected.
- They can be controlled by bush clearing, spraying and encouraging dense farming settlement. Fifteen million sq km of fertile land in Africa could be farmed if it were free of tsetses.
- Ingenious research has produced some solutions: the *sterile male technique* where sterilised male flies are released to mate with the females; the *tsetse trap* using a chemical bait which smells like the breath of cattle to attract flies into a net.

Bilharzia snails
- Water snails carry bilharzia schistosomes. The snails are the hosts for minute parasite worms which bore through human skin and lay eggs in different parts of the body. This causes damage to kidneys, gut, eyes, even the brain.
- Two hundred million people in the world suffer from this parasite.
- Killing snails is important but strict sanitary rules are also necessary. Eggs are passed on when people urinate into the general water supply. People, especially children, need to keep out of the infected water and get their drinking water from safe places.

Figure 19.5

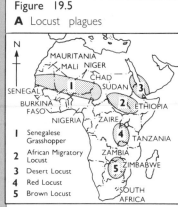

A Locust plagues

1 Senegalese Grasshopper
2 African Migratory Locust
3 Desert Locust
4 Red Locust
5 Brown Locust

B Yellow fever and malaria

Areas liable to:
Yellow Fever and Malaria
Malaria

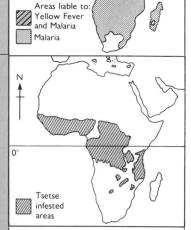

Tsetse infested areas

C Tsetse fly

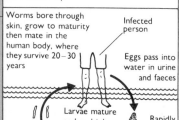

D The bilharzia infection cycle

1 All the pests or diseases listed have a geographical distribution related to environmental conditions.
2 All have a *cycle* of stages of infection (see bilharzia)
3 Many need a 'host', that is, somewhere to live while they develop (a human, an animal or a snail).
4 There has been much ingenious research (see tsetse fly).

In the gloom about diseases and pests in Africa there is a bright spot. Research shows that there are over 30 Western diseases that have not reached high levels in Africa, although they do exist. They include some cancers, asthma, constipation, heart diseases, multiple sclerosis, migraine – many of which may be linked to nutrition. The normal diet of rural people contains a high proportion of natural foods including vegetables, grains and fruit with plenty of fibre. They eat smaller amounts of animal fats and highly refined foods like white sugar, white bread and white rice. Unfortunately many families, especially in towns, now eat more of these refined foods.

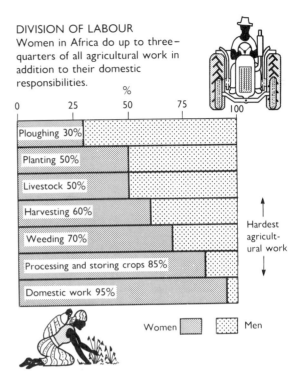

Figure 19.6 Women's share of the work

Wasted lives, wasted resources

It is difficult for ordinary people to avoid diseases related to the environment in which they live. There are 2 other ways in which people's lives can be wasted:
• If they become refugees, with no home, no country to belong to, no land, often even no family, living off government relief.
• If they are women being under-used and prevented from carrying out their role in the development of the country.

There are millions of refugees in Africa (see Chapter 23) for a number of different reasons: drought, wars, tribal divisions, colonial carving up, political disputes.

A new life can work if people want it to

About 165,000 Saharwi people fled from Moroccan occupied Western Sahara in 1975 to four refugee camps in Algeria.
• They were allowed to settle in a desert region thought to be uninhabitable with no natural vegetation and little underground water.
• International aid provided cloth for tents and clothing, food supplies and tools.
• The people created their own survival programme.
• Each camp has its own agricultural project with deep wells, irrigation, and soils collected from elsewhere to grow vegetables and fruit. When rain fell they planted wheat. Camels, goats and even some cows, now provide milk for nursing mothers and young children.
• They have set up schools and hospitals with teaching facilities and women's committees provide local health and social care.

These refugees have strong leadership and a long history and culture. They support the idea that *anything* can be made to work in Africa, if people want it to.

Women: over-used yet under-used

1 Women in Africa:
• do up to 75 per cent of the farm work including the exhausting work of weeding and harvesting
• prepare all the food (grinding grain, pounding yams, pressing oil)
• cook it
• carry water
• carry wood for fuel
• bear children (9 months of pregnancy, then

breastfeeding)

Figure 19.6 shows the traditional division of work between men and women.

2 Food farming is traditionally the domain of women. Perhaps because women were seen as the main support of the family, a woman was often given land by her mother at birth.

3 Colonial authorities customarily dealt with the men:
- agricultural officers gave advice to men
- the head man was the chief contact
- land titles were given to men
- banks lent money to men
- government grants often helped with men's work: bush clearing, marketing
- governments encouraged cash crops grown by men rather than the food crops grown by women

4 So women often lost their traditional land rights and the incentive of ownership. Check back to page 81, the Gambia.

5 Men are often away working in the mines, commercial farms or towns. Women bear the responsibility for keeping family and community life going and doing the jobs men used to do.

6 Africa's food production is falling. Importing expensive food cripples a country. It is essential to increase the production of food crops. This depends on the cooperation of women and they need all the support they can get.
- With access to advice, training and loans of money they could do much more.
- If their work is made easier by provision of simple things like wells for water and mills for grinding grain and pressing oil, they can work much more productively at their farming.
- They could also do more for their communities, if they were given responsibility and help.
- They, more than anybody, can help to keep population growth in check if encouraged and supported.
- This is why we said at the beginning that women are under-used.

'People particularly women, should share in decisions not just about the food they grow, but about the number of children they choose to have.'

At the Nairobi conference at the close of the United Nations Decade for Women in 1985 it was obvious that African women were not seeking confrontation with their men, but saw their role as complementary. They see the traditional values of family life and social togetherness as great strengths in the face of economic hardship.

Figure 19.7 Improved technology: women pedal-threshing rice in the Gambia. The women of 40 villages now share in Land Credits for improved rice production, saving the cost of imported rice. The IFAD rice project embanked the Gambia River swamps, work too heavy for women. Re-read page 81

Chapter 20 Pressure on the land

> **Key words**
>
> Security, rural tradition, desertification, degradation, sustainable development, integrated rural development, tree crops, agroforestry, fish farming

So far we have been concerned with the first resource: people, and how to ensure that they have a chance to live healthy and worthwhile lives. Other resources are the land, including minerals and power supplies, and the transport network which promotes trade and the exchange of ideas as well as goods.

How do people feel about the land?

Ownership of a piece of land gives security in Africa. But not everyone views it in the same way.

To find out how you feel match the 'heads' and 'tails' in the following statements. Take the first statement on the left-hand side and look down the right-hand side for the correct ending. Write the whole sentence down in your jotter. Do it for each one in turn. Check first, to make sure you can use up all the tails.

questions are sometimes set this way.

Rural life in Africa

Village communities develop a way of life that suits their own particular area and tradition.
- Each has a stable and well-tried life-style that expresses a balance between social values (family, ways of bringing up children, each person's place in the community); and the physical environment of weather, soils, and the kind of land farmed and its availability.
- The 'technology', that is the tools and the kind of transport used, meets their needs and supports a proven way of life. Traditional farming maintains soil productivity and the bush or forest cover.
- This strong rural tradition persists despite the immense variety of farm systems, and the many outside influences that exist.
- The human and ecological balance continues until some kind of pressure, like years of drought, or too many mouths to feed, triggers a change. The following 'stories' describe some of the ways this could happen.

Story A:
- The government builds clinics, provides doctors to improve health.

Result: fewer babies die so more mouths to feed.

1 When population densities are low no one minds if a family takes and clears new land for farming...	...to small scattered pieces of land	A
2 When land is divided between a man's children for several generations...	...because there is plenty of land for all.	B
3 Governments want to apply modern farming techniques...	...the pieces gradually become smaller and smaller and a man's land is scattered.	C
4 It is difficult to apply large-scale farming techniques (ploughing, etc.)...	...because they are afraid of losing their right to the land.	D
5 When land is scarce...	...in order to produce more.	E
6 Farmers are afraid to let the government regroup (consolidate) their land...	...most people want to own a piece however small.	F

The answers are 1B, 2C, 3E, 4A, 5F, 6D. Which of the questions and answers are the most important for your country? Are there others you think should be added? Note that examination

- Families have difficulty in growing more food

Result: able bodied men decide to go to work in mines, or on commercial farms.
- Some village work is too heavy for women

and young children, land is neglected.
Result: food crops suffer, yields drop.
Story B:
- The government builds primary schools, youngsters get formal education.
Result: teenagers want to earn money, take jobs in town.
- They say to other young people: 'Farming is hard, hot, dusty work. Life in town is exciting.' They leave home too.
Result: farms neglected, crops yields fall.
Story C:
'The rains didn't come last April. . .
Write the rest of it yourself.

The result in all the cases is the same. The land suffers, farm production drops, there is less food and cash, and the poor harvest puts families at risk. Figure 20.1 shows how food production has been dropping in Africa relative to the population which needs to be fed.

Famine and poverty in Africa

People are sometimes faced with terrible disasters like floods, a volcanic eruption, or a locust swarm. But long-term disasters nearly always result from several things that combine to make a bad situation worse. Sometimes the situation is aggravated by the actions of people, or even by governments. In the two 'stories' on page 175 and 176 the government was trying to make things better: they were shocked to find out what actually happened. The lesson is, that the different aspects of 'development' have to take account of each other and keep in step. The impact of a new development may be more complex than it first seems. This is why integrated development is so important.

When the rains fall: Africa's desperate need for water

There is no doubt that drought is one of the greatest hazards to crop production.

Africa has always suffered years when the rain either failed or was below average. But there were fewer people then and farm effort went into food not export crops. For 5 years from 1968 to 1973 the savanna zone of Africa suffered severe drought. It was especially bad in the Sahel (see Chapter 8) and was the cause of migration to find water, food and work. Unfortunately the 1970s and 1980s have seen more dry years. Drought was even worse in 1975, continued through 1977 and became entrenched in 1983–84 (see Figure 20.2). However some people say that

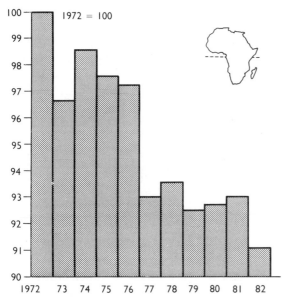

Figure 20.1   Africa's lost food production. This bar graph shows how the amount of food produced in Africa has shrunk *per head* since 1972. Note that the scale on the left side of the table is an *index number*. Food production per head in 1972 has been given the value 100, and the following years are measured in relation to this. The smaller size of the bars does *not* mean that less food has been produced, but that the increase in food production has been swallowed up by the greatly increased numbers of people who eat it.

Figure 20.2   The victims of years of drought in the Sahel. These starved cattle died as a result of drinking water on empty stomachs.

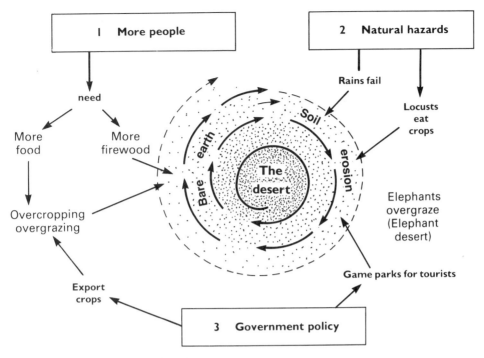

Figure 20.3 Some reasons for the expanding desert

Africa's problems do not come from drought alone, but from the over-used, impoverished soils, low-yielding crops, low prices paid to farmers for food crops, and the destruction of the environment that we call desertification.

The expanding desert

Figure 20.3 shows how many different factors have combined to create the present disastrous situation. There are many elements in the process and they are all linked. Take, for example, *firewood*:
- To cook their food (if they have any) people hack down trees and bush for firewood, or use plant stems and animal dung that should be used to fertilise crops.
- Gradually the ground cover of trees, bush and small plants disappears. The relentless sun bakes the exposed earth hard, and destroys its structure. The next rainstorm carries off the top soil and starts sheet or gully erosion. Both water and soil are wasted (see Figures 20.4 and 20.5).
- The damage to the land is permanent: it is *degraded*. All these things reduce the productivity of the land and its ability to support the present population. It is made worse if more and more people have to be fed.

But it is not only the crop farmers who are at risk. Herders have always moved to better pastures with the rains. In rainy years both the settled field-crop farmers and pastoralists extend their farms or herds. The government encourages the growing of commercial crops to increase revenue.

But when the rains fail the pastoralist's search for grazing and water brings them face to face with the settled villagers. Both are competing for scarce resources. There can easily be conflict along the interface where the two communities meet. This is especially so if farmers are struggling to keep their own work animals alive.

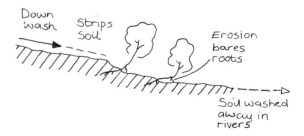

Figure 20.4 Soil erosion exposes tree roots – look at 20.5

Figure 20.5 Gully erosion in northern Somalia. Occasional intense storms strip the soil from the bare ground where overgrazing has removed the vegetation. Deep gullies are formed, undermining the few remaining trees

same level. The agricultural officer helped by bringing a transparent tube, half filled with water, which showed when the line of stones was level!
• A second line of stones was built to make a second terrace, and then a third (see photograph, Figure 20.6).
• When the rains came the stones slowed the run-off and trapped soil. Small plants grew among the rocks and gradually became a barrier to soil erosion. Next season the food crops on the terraces were better because the soil was a little deeper and the rain sank in instead of rushing away.

In Ethiopia the terraces were made with tractors by the government agency. Two years later they had disappeared because the local farmers 'did not know what they were for' and had found other uses for the stones.

It follows that action on several fronts is needed, and that really lasting solutions can only come from integrated rural development. 'Food aid' is short-term help for people in desperate situations. Lasting improvement is only possible if sustainable forms of development are adopted, that is, methods that will carry on into the future. The hardest decision for people and governments is to know where to put scarce resources.

The drive for more food

The farming technology of developed countries has rarely been a success in tropical Africa. The success stories come more often from adapting local methods in what is often called *appropriate technology*. Simple methods are easier to understand and to communicate.

A good example of self-help at the grass-roots level comes from Burkina.
• Everyone, including old people and children shared the work of carrying stones, to build low 'bunds' (walls) along the contour, keeping to the

Figure 20.6 In Burkina Faso stone bunds are built across the hillside following the contour to retain water and reduce run off and soil erosion. This is a form of community self-help but most of the work is done by women and children

Field and tree crops

There are many plants which can survive in harsh environments. Two of these are the Balanites tree in the Sudanese desert and the Yeheb of Somalia. Both are at risk, dying out because they are cut for firewood or the last fruits and seeds are eaten by hungry people.

It they are harvested, that is, the branches are cut but the trees survive and some nuts are saved for planting, they could be a very productive crop.

> 'Development cannot be sustained without conservation: we must make the deserts bloom!'

What is needed is a planting and protection programme to conserve the native plants.

Agro-forestry

Many of the older farm systems of Africa were forms of agro-forestry. Farmers cleared the understorey in the forest or bush but left large trees. They planted ground crops such as cocoyams, started cocoa farms, and harvested oil palms and other tree crops. In tropical forests it works within the ecosystem and can be a form of multi-storey cropping. It can also be used in oases (see Tougghurt, page 104). It is a management system that sustains soil fertility, needs no expensive inputs, and increases the overall yield of an area of land.

Successful stock rearing

The International Livestock Centre for Africa does research and tries out projects in several countries including Botswana, Mali, Kenya and Nigeria.
- It has found that some breeds of small African cattle, sheep and goats have developed a natural immunity to the cattle disease, nagana, carried by tsetse flies. They also give good milk yields.
- Sahel pastoralists are being encouraged to increase camel herds. Camels survive droughts that kill other animals, and continue to give better milk yields.

Figure 20.7 Children make long walks to collect firewood often reducing their time in school

- Few families have two oxen for ploughing. A single ox plough has been designed costing only $5 and it can be made by village blacksmiths.

Fuel for cooking

On page 177 cutting wood for fuel was said to be a key element in desertification. In some countries firewood is being used up 10 times faster than trees can grow. This affects soil erosion, soil fertility and water supply. What are the solutions for towns as well as for the countryside?
- Fuel wood plantations are needed. Large-scale planting must be protected from animals and cutting at first. Then wood would be harvested, that is cut in a controlled sequence, everyone getting a share.
- More efficient cooking stoves would save fuel.
- In some places coal bricks could be used (see page 134).

Such developments would save much time and hard work carrying wood long distances (see

Figure 20.8 A fishing 'company' on the Ghana coast

Figure 20.7); they would save the natural vegetation cover; and they would allow dung to be used to fertilise crops.

Fishing and fish farming

Fishing is one of the widespread activities in both coastal and inland areas in Africa (see Figure 20.8) and fish are a very important source of protein. The table shows fish catches for 1981. Recent developments in coastal fisheries include outboard motors for traditional canoes, larger powered boats and commercial drying ovens. Dried fish are traded to inland areas and towns.

> 'Give a man a fish and feed him for a day. Teach him how to fish and he'll feed himself for a lifetime.'

Coastal fisheries and fishing in lakes and rivers are important but fish farming could provide more of this important food. Lakes and fish ponds could be stocked with fish. Small and large dams in dry areas could also be stocked, and fish ponds built to use surplus irrigation water. Swampy areas near rivers or in badly drained land could be converted to fish ponds. Pond management is needed if such conversion is to be successful, and this needs knowledge and care. Some indigenous fish can be very successfully farmed, such as tilapia (chambo) in Malawi, but there can be disasters too. When the Nile perch was introduced into Lake Victoria, there was a drastic drop in the numbers and species of other fish, which it ate!

Fish catches, 1981	000s metric tons
South Africa	619
Nigeria	496
Morocco	382
Namibia	250
Ghana	240
Tanzania	226
Senegal	206
Uganda	167
Egypt	142
Angola	124

1. Add up the totals for different parts of the coast, for example, Nigeria, Ghana and Senegal for West Africa.
2. Where is most of Uganda's fish likely to be caught?

Combining small and large-scale farming

Figure 1.3 in Chapter 1, page 10, shows very simply how the two main sectors of farm production relate to each other. One circle, the commercial farming sector, is concerned with market crops. The other circle represents domestic crop production. In the centre the two overlap. Most farmers have to try to grow food for themselves and grow a market or an export crop. This diagram is true for both the woman farmer who sells surplus yams in the local market, or an outgrower who sends sugar cane to the estate mill. In each case food crops are an essential part of the system.

Look back at Chapter 6 to see how this works for a small farmer in an outgrower system. Sugar, tea and coffee can be grown by small farmers on the edges of large estates where they can get benefits from a guaranteed market, and help with advice and equipment. They get a cash income but are still independent and have their own land.

The right kind of food

Growing enough food so that people do not feel hungry is an important goal. An equally important one is to provide the right kind of food so that people are healthy and full of energy.

There are 3 main forms of food: carbohydrate, fat and protein. They all provide energy but only protein supplies 2 things that are essential for proper growth: nitrogen and amino-acids. Protein deficiency in young children means that growth is almost at a standstill. A severe lack of protein leads to break down in the liver and other organs and to chronic ill-health. Many children under 5 years old in Africa suffer from a protein deficiency disease, kwashiorkor. Young African children are also likely to suffer from diarrhoea and lose the value of their food as well as catch other diseases more easily than children elsewhere.

Protein is found in animal products, including milk, fish, and in some vegetables, particularly peas and beans, and in cereals and groundnuts.

Changing food habits

Every part of Africa has its own staple food crops which most people can produce on their own farms (see the fact box on page 182). Any surplus can be sold in the local market. Most villages brew a local beer that costs very little. But eating habits are changing. This is because people are moving from the countryside to work in towns, People from Europe, the Middle East and India have come to work or settle. Changing food habits are part of social and cultural change.

Many people are eating bread made with wheat flour (see Figure 20.9). Few African countries grow wheat so imports have to be paid for. More African countries could grow wheat as a cool season crop. Tinned foods and bread can be bought even in small villages. Tinned foods, Coca-Cola, beer and many other things are tasty, convenient, and also have status value. Most of the staple foods take long hours of work pounding or grinding each day.

Figure 20.9 A city poster encourages people to buy bread. Note the varieties of housing and the 'informal sector' tyre fitting shop

Facts: Staple Foods

Crop	Climate and Cultivation temperature	rainfall	Altitude	Soils	Comments
Maize: annual cereal	30°C daytime growing temperature no frost	600–1,200 mm annually	0–2,000 m	Well-drained, wide variety	Many varieties; high yields particularly from hybrid seeds
Sorghum: (giant millet)	25°C needed in growing season	400–1,000 mm annually	900–1,500 m	Heavy, well-drained soils	Often grown with millets, but lower in protein
Millets: Finger millet	Many varieties suiting different conditions. Tolerates lower temperatures 18–27°C	800–1,200 mm annually	Two main ones are: 900–2,400 m Higher altitude	Tolerates poor soils	Can be stored for long periods (up to 10 years) Seldom eaten by birds.
Bulrush millet	20–25°C	250 mm minimum. Very drought resistant		Fertile or sandy soils	Very nutritious
Others	Many produce good yields in difficult conditions. Some semi-wild varieties are very important in areas of unreliable rainfall				
Bananas: tree crop	27°C mean, shelter from wind	900 mm annually evenly spread	0–2,000 m	Deep, well-drained soils	Fruits picked after about 3 years. Intercropped. Many varieties, some for beer. Low in protein
Cassava: root crop (tuber)	High temps	Tolerates very dry conditions	0–2,000 m	Tolerates poor soils	Easy to grow. High yields. Can be stored in ground for many years. Important reserve crop but low in food value
Yams: root crop	20–30°C	1,000–1,500 mm	0–1,500 m	Fertile soils	Low in protein but better than cassava. Easy to grow mixed with other crops
Rice: annual cereal	High temperatures and good sunlight	Seasonally flooded land or irrigation	0–1,200 m	Heavy, water-logged	May need levelling and 'bunding' to flood and then drain fields. Matures quickly
Wheat: annual cereal	18–20°C, 25°C maximum	500–900 mm	Grown at high levels	Well-drained loamy soils	Needs a dry period. Can be irrigated in cooler, dryer areas, or grown as winter crop

These staple foods provide most of the carbohydrate that people need to stay alive. Some of them are very low in protein and children who are fed on these alone often suffer from malnutrition. However, many different pulses (groundnuts, peas and beans), vegetables and fruits are also grown and these provide vital proteins and vitamins.

Chapter 21 Problems of expanding cities

> **Key words**
>
> 'Bursting' cities, urbanisation, unofficial housing, on-site improvement, new capital cities

URBAN GEOGRAPHY INDEX	Page numbers
Agricultural towns in west Africa	82–83
Oasis towns in north Africa	102–103
Duka towns in east Africa	54–57
City life in north Africa	95, 100
Contrasts between town and country	183
Urbanisation: how towns grow	54–55, 184–185
New capital cities	85–86, 187–188
On-site improvements in Lusaka	186–187
Mopti, a river town in Mali	76–77
Mombasa: a study of a port	56–59
Nkana–Kitwe: a study of a mining town	128, 130–131
Nairobi: the expanding city	185–186
Johannesburg: transformation of a city	149–150
Lagos: a multi-functional metropolis	83–85
Cairo: the largest city in Africa	114–115

Present-day African cities fall into 2 main groups:
- Long-established cities, including the Islamic towns of the northern half of the continent.
- Former colonial cities in countries which have now been independent for at least 20 years. They also show some typically African forms of urban development. For example:
  – ethnic and kin-group neighbourhoods
  – a strong and continuing rural-urban link.

Both groups are 'bursting' with the surge of incomers. In addition, innumerable small and medium-sized towns and service centres have grown up throughout the continent. This process of urbanisation means that more of a country's people now live in towns.

Many writers stress the varied character of towns in Africa. This chapter deals with:
- contrasts between town and country
- the move to the towns
- urbanisation and its problems
- some solutions: self-help housing and new capital cities

Use the signpost table, to refer back to urban studies throughout the book.

Contrasts between town and country

Many ordinary African families now have amenities similar to those of families in rural areas or small towns in Europe. Well-off people have cars, domestic help, a good education.

There are two main exceptions:
1 Families living far from towns have to rely more on what they and their neighbours produce. The quality of family life in smaller communities should not be undervalued. However,
- The lack of a good water supply and electricity can be a problem. Without good water health is at risk; and a long walk to fetch water is a great waste of time that could be spent producing more food or going to school.
- The lack of electric light or power limits study and all kinds of small businesses.

2 Some parts of towns become overcrowded when country people go there looking for work. It is natural to stay with a family from your own village; but if a whole family moves, then they have to build a temporary home and re-create village life in the town suburb. There is usually no proper water supply or electric light. That is why planners are putting in basic services like drainage, water taps and roads in some areas where people can come to build their own homes (see Figures 21.3 and 21.4).

The move to the towns

Pressure on the land (see Chapter 20) is one of the reasons why people leave the countryside and go to the towns. In Britain 100 years ago about three-quarters of the people lived in the countryside and only one-quarter in the towns. Now the opposite is true. In Africa there is a large rural population. But the towns are growing at such a rate that by the year 2,000 the situation may be very different. The table on page 184 shows how big some African cities are.

When towns become very large they develop problems:
- bringing in enough food every day

African cities:	Population in millions, 1983/84		
Cairo	12.0	Casablanca	1.4
Lagos	5.0	Addis Ababa	1.4
Alexandria	3.0	Khartoum/Omdurman	1.3
Kinshasa	2.5	Abidjan	1.2
Soweto	2.0	Luanda	1.2
Johannesburg	1.8	Nairobi	1.1
Cape Town	1.8	Durban	1.0
Algiers	1.7	Tripoli	1.0

- congestion when more traffic uses roads built for few cars or lorries
- providing a water supply
- getting rid of rubbish and sewage
- providing homes for newcomers.

For the last 40 years there has been a persistent drift to towns. There is a marked increase in the numbers of people unemployed or only partly employed there.

How serious is the situation?

1 There are over 550 million people living in the African continent.
2 The total population increase for Africa is around 3 per cent per year.
3 This means there are about 15 million extra people a year.
4 But the urban growth rates are 7–10 per cent per year, for example:
- Cairo from 2 million in 1947 to 12 million in 40 years.
- Abidjan from 119,000 to 1,250,000 in 30 years.
- Dakar from 132,000 in 1945 to nearly 1 million now.
- Gaborone from under 5,000 to 60,000 in 20 years.
- Lagos from 300,000 in 1968 to 5 million today.

5 There are more young people, for example:
- In 1970, 47 per cent of the total population (345 million) was under 15 years old. How many million is that?
- In 1983, 48 per cent of 530 million was under 15 (... million?).

6 A totally unemployed person is a drain on his family or community whether in town or country.
7 It is important to develop labour intensive and informal work opportunities. However, newly made jobs are often not enough to cope with the increasing population and more school leavers.
8 Rural development schemes must be seen to work. When the advantages of living in rural areas are appreciated the towns might have a chance.
9 Migration to nearby countries sometimes helps the unemployment problem but when jobs are scarce there too, foreign workers are sent back.

Towns only remain safe places to live in if they are well organised and well planned. It is difficult to build well when the situation is already unmanageable. Planners make their studies in order to find out why towns grow or what causes them to grow in particular ways. This helps them to forecast what is likely to happen in 10 or 20 years' time.

Urbanisation: how towns grow

An urban centre is a place where a lot of people are clustered together. In towns, most of the people earn their living from trade (shops, hotel keeping), industry, or transport, instead of from farming. We give towns labels like market centre, port, capital city, or mining town, because of their main function. But many different activities

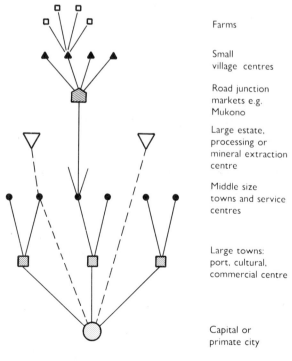

Figure 21.1 A hierarchy or rank order of urban places. Make one for your own area

go on, and most places, both large and small are multi-functional. Check back to Mombasa, page 58, for the meaning of function.

Towns and cities also provide different levels of services. For instance, a small local centre may have one to two shops and a petrol station, while a capital city has a whole range of administrative and commercial services to offer. Figure 21.1 shows a hierarchy, that is, a size order, of towns and how they relate to each other.

The term urbanisation is used to describe the process of small settlements growing into urban centres. Although in parts of Africa there is a long tradition of urban life, generally the level of urbanisation is low. But the rate at which towns and cities are growing is very high.

Nairobi, the expanding city

Nairobi is one of the most flourishing cities in Africa. It is a 'new' town, only as old as this century. It began as a railway camp and stores-point when the line from Mombasa was built to Kisumu on Lake Victoria at the turn of the century. At Nairobi, the very difficult section of the line crossing the Rift Valley begins.

Figure 21.2 shows the core area of the city, the older city boundary dating from the 1920s and the present boundary which goes well beyond the built-up area.

Year	Population of Nairobi
1963	343,000
1969	480,000
1984	1,100,000
2000	3,000,000 (projected)

Nairobi is faced with formidable housing problems. Every year 10,000 homes are put up, but 80 per cent of them are 'illegal'. About 40 per cent of the population of Nairobi live in what has been called an alternative society. A 1970s description said:

'Homes are made of boxes, plastic bags, flattened tin cans, with no water, no lavatories, no refuse collection... One-third of Nairobi's workers are employed in the informal sector: carpenters, cobblers, tinsmiths, bicycle repairers, photograph frame makers, working almost entirely with waste material thrown out by the other Nairobi.'

This is the 'self-help' city that tourists do not see. It grew because the formal city could not provide

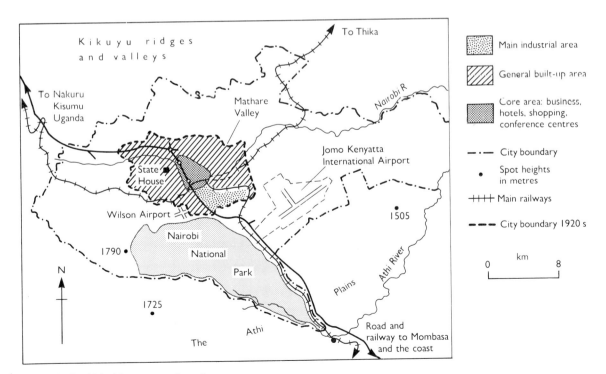

Figure 21.2 Nairobi: an expanding city

186 The continental view

enough jobs, houses or schools for the migrants from the countryside. The problem for Nairobi is not as acute as in some other African cities (Cairo, Lagos, Lusaka, Rabat) and the jobs and services are a positive aspect of informal living.

Unofficial housing: an invaluable form of self-help

The term 'squatter' was used in Australia for people who built homes on empty land. It has

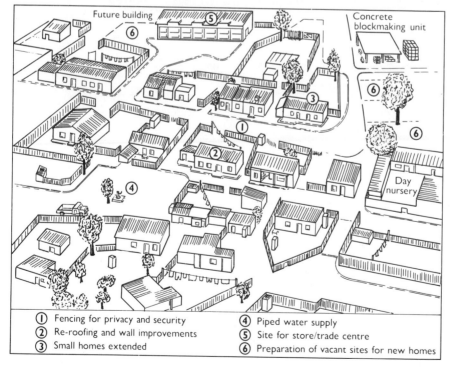

Figure 21.3 & 21.4 How to use community resources and effort to improve a neighbourhood. The photograph shows an informal housing area; the line drawing shows the improvements that people can make, even with limited resources, if there is security of ownership

① Fencing for privacy and security
② Re-roofing and wall improvements
③ Small homes extended
④ Piped water supply
⑤ Site for store/trade centre
⑥ Preparation of vacant sites for new homes

come to mean people who take over an unused building or put up a home on someone else's land. Squatter settlements have many other names: shanties, informal, unplanned, unofficial improvised. But they are often urban villages with strong community spirit, combining self-help and mutual aid. The people show immense initiative and a great deal of hard work.

Governments faced with the impossible task of building homes for an upsurge in the birth rate, or an in-surge of people from the countryside, now make use of the do-it-yourself movement and provide basic amenities on planned sites.

Two of many types of 'self-help' schemes are funded by the World Bank.
1 Site and service schemes build 'house-cores' on new estates. The core is a lavatory and kitchen (supplied with water) and householders build the rest.
2 On-site improvement helps to improve squatter communities without moving them away.

On-site improvement in Zambia

An example of the second policy comes from Lusaka, Zambia. The following describes first, how this was organised, and second, what happened to one of the mining townships near the copperbelt town of Mufulira.

Figure 21.3 is an aerial photograph of informal housing. The lay-out is orderly but there are no amenities. Figure 21.4 shows how it has been made a more comfortable place to live in.

Use the key to check the following:
1 How has the water supply been improved?
2 How has safety and privacy been increased?
3 Has (5) been built yet?
4 Work out how many *extra* rooms have been built
5 How many homes now have pit latrines?

What is not shown in the line drawing is that:
- there are more shade trees and gardens are planted with tomatoes, pumpkins, maize and flowers
- there are garbage pits
- a ditch takes surplus rain water away
- there are stalls on site
- the group of 20–25 homes have a committee to organise self-help and expel trouble-makers.

While copper was earning hard cash on world markets Zambia was one of the front runners in Africa for housing improvement. The drop in copper prices forced Zambia to make drastic economies. This is what happened to Kawana West, a township near the copperbelt town of Mufulira.

Kawana means 'a small beautiful place' in the Bemba language. In 1970, when 40 families were resettled, there were 4 water taps. In 1975, when copper prices fell, the money for upgrading of the unplanned settlements ran out. By the 1980s Kawana had 12,000 people and the original 5 townships had become 38. There were still only 4 water taps. In Kawana about 80 per cent of the work force aged between 15–30 years were unemployed. The families were kept alive by money sent back from those in work; by collecting firewood and making charcoal; by working as nightwatchmen and selling single items, like cigarettes, from larger packets. The Zambian government subsidised maize flour (mealies) so that it was half of cost price.

Because Zambia was unable to repay loans, the IMF proposed economies, one of which was the removal of the maize subsidy. This led to food riots. But other cuts reduced the amount spent on health, schools, transport, and caused job losses on the mines. Now people have to 'decide whether to eat once a day or eat porridge only every other day, so as to buy the children the shoes they need to get into school'.

'It is the very poor who suffer most.'

The increasing size of all African cities brings with it problems similar to those described here. A city almost at a standstill was one of the reasons for creating a new purpose-built federal capital at Abuja (see Lagos, pages 85–86). Other countries than Nigeria have done this too.

New capital cities

During the colonial period most countries were administered from the place that was most convenient for the colonial power.
- This was often a port, and became the major administrative, business and industrial centre for the country.
- The combination of several major functions caused congestion.

188 The continental view

Capital Country	Lusaka, Zambia	Nouakchott, Mauritania	Gaborone, Botswana	Lilongwe, Malawi	Dodoma, Tanzania	Abuja, Nigeria
Year planning began	1931	1960	1964	1968	1972	1976
Population	575,000	600,000*	80,000	120,000	100,000	25,000
Central location	Yes	No	No	Yes	Yes	Yes
Previous capital	Livingstone	St Louis (Senegal)	Mafeking (South Africa)	Zomba	Dar es Salaam**	Lagos**

Notes: * includes large refugee population   *a new capital city for Ivory Coast is being built at
       ** parliament still in old capital   Yamoussoukso

- The location at one end of the country made it difficult for people to reach the capital from other parts.

The above table gives brief details of several new African capital cities that have mainly been started since independence. They are seen as an expression of independence and national unity, especially if they are built in an area not dominated by a major ethnic group. In nearly all the cases, the new location is much closer to the centre of the country, and will help the economy of the area by bringing in jobs and public expenditure.

Gaborone: a good working example

Before independence, Botswana was administered from Mafeking (now Mafikeng), in South Africa. The location of Gaborone was chosen because:
- it was not too far from Mafeking (although it is not central for Botswana)
- most people live in the eastern part of the country
- there was a good water supply and land for building
- it was on the railway line, the main transport artery
- much of the land was in public ownership.

The plan for the new capital, Figure 21.5, has provided a central shopping area based on a car-free pedestrian mall; parliament and government buildings; a university; pleasant residential areas offering house plots of different sizes; and industrial zones close to the railway but separated from the housing areas.

There has been very little unplanned housing in Gaborone, and what there is has been improved by self-help building loans to the residents, and the provision of water taps, better roads, and latrines. A new international airport has been built to the north.

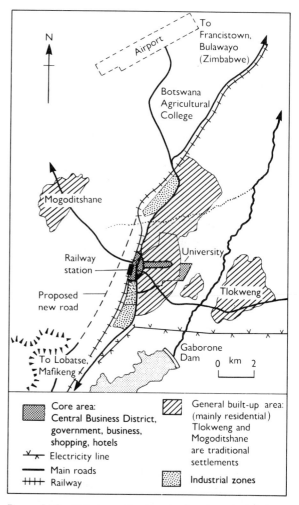

Figure 21.5 Gaborone: the layout of a planned capital

Chapter 22 The development of industry and infrastructure

> **Key words**
>
> Primary, secondary, and tertiary sectors, informal sector, support system, transport planning, geothermal energy, appropriate technology

This chapter is about:
- the basic needs for industrial development
- infrastructure: the essential support system

Some of the largest enterprises in Africa are *primary* industries related to minerals or processing primary or 'raw' agricultural products like sugar. They are found at the mineral or agricultural site, often far away from other centres.

Secondary industries, like assembling motorcycles or brewing, are more often placed in towns because they use a range of materials brought in from several places and must have good access to customers. Towns provide a market and have large numbers of potential workers.

Industrial or commercial enterprises need support services of all kinds. These are the *tertiary* 'industries'. They employ people who range from typists and bar keepers to fitters who repair cars and tractors, and include computer operators and government clerks.

Most countries have large import bills for manufactured goods. If some of these can be made at home scarce foreign exchange can be saved and used to buy others that countries cannot make for themselves. Industries can also help to add value to primary products for export. For example, plywood earns much more than logs.

The basic needs for industrial development

However large or small industries are, they depend on the following:
1. The supply of raw materials.
2. The demand or market for goods.
3. Power supplies.
4. Workers, who form the labour force.
5. Capital, that is, the money to finance development.
6. Management to take decisions and see that everything works well and keeps going.
7. The support system of roads, railways, power stations, construction industries, training schemes, housing, etc., called the infrastructure.

1 The supply of raw materials

An industry operates cheaply if it has easy access to raw materials.
- Industries based on bulky raw materials which contain much waste, like copper ore and other minerals, are usually located near the source of their raw materials.
- Many agricultural products are also processed at their site, for example, sugar at Mumias. Timber and plywood processing at Sapele is an example of both primary and secondary processing near where the trees grow.
- Where several raw materials are required from different places it is better to site an industry at a convenient centre with good communications. This is the way 'corridors' of industrial development grow up like the PWV triangle in the Republic of South Africa.
- Water is an important raw material in manufacturing industry.
- The raw materials for the tourist industry are the things visitors want and are willing to pay for: sunshine and beaches, spectacular mountains and rift valleys, game animals that are rare in other parts of the world.

2 The demand or market for goods

Goods must be sold at home or abroad if manufacturers are to earn a living.
There are 3 possible markets for African industries:
- The home market (within your own country). A well-populated country like Nigeria can have a large home market if people have enough money to spend.
- Other neighbouring African countries. Many African countries produce similar products: sugar, coffee, timber, diamonds, which they cannot sell to each other.

- The world market. Fluctuations in world market prices have drastic effects on exports and competition is fierce.

3 Power supplies
Human hands, animals, wind, flowing water, and fuels such as wood, coal and mineral oil have all been used to create power. Electricity is a very flexible type of power if there is a system to distribute it from the generating plant. Electricity can be generated from fuels (see coal, page 140), water (see page 199), and geothermal sources (see page 194). Africa has huge reserves of water power.

4 The supply of workers
The term labour is used in industry to cover a varied workforce. It includes estate and forestry, factory and transport workers and all those in tertiary jobs: shop and café keepers, drivers, office staff and typists, teachers, accountants and computer analysts.

5 Capital: the money to invest in development
Even starting up a market stall needs cash. A larger industry requires capital to pay for a factory or workshop, buy supplies of materials, pay workers, and 'overheads' such as rent, telephone bills, electricity bills and advertising. This is long before any money comes back in from selling the product. Money comes from many different sources: national banks, the government, the World Bank, and private investment from individuals and from firms overseas, many of them multinational.

6 Management
Management is really a form of leadership, an organisation skill. Some people are quick to see possibilities and take responsibility and initiative. They can be trained to have technical or commercial know-how which helps them to organise workers and stages of production and marketing. Every country needs a combination of small and large-scale businesses. Small-scale businesses give people a chance to learn both work and management skills.

7 Infrastructure: the back up to industry
People now realise that successful industries depend on an interlocking support system. This must be available before a development scheme begins. If there are no roads the raw materials will not arrive and the products cannot be taken to markets. The public utilities must be there to supply water and power. Workers need housing and medical care. Infrastructure is discussed more fully later in the chapter. Figure 22.2, the industrial wheel, shows just how complicated industrial development can be. It is easier to understand how it works by first looking at a real life example – a tourist hotel by Lake Malawi.

A tourist hotel by Lake Malawi
Figure 22.1 is a diagram of a tourist hotel on the shores of Lake Malawi.
1 The lists below the drawing show groups of needs: the transport group, the 'good food' group, the 'comfort' group, the leisure activities group
2 These lists under the drawing are not complete and would be different for other tourist resorts.
3 Can you suggest what they might be for a tour on the River Nile in Egypt to see antiquities, a climbing expedition to Mount Kilimanjaro, a game viewing safari to Hwange, Zimbabwe; a beach holiday in Tunisia?

Tourism, an 'invisible' industry
Tourism is an 'invisible' export industry. It does not appear in the export figures, but it earns millions of dollars of foreign exchange for many countries. For example, it is the 2nd or 3rd foreign exchange earner for Kenya in most years. It also employs thousands of people.

Money received from tourists in 1982

	million $US
Morocco	425
Egypt	386
Tunisia	370
Kenya	185
Algeria	167

The money earned by this 'invisible' export for Morocco in 1982 was equivalent to 20 per cent of the *visible* export earnings. Check back to the trade summaries for the other countries to see how these figures compare with their visible export totals.

The system known as 'package holidays' became popular in the 1970s. This brought many people on short visits and 'camera safaris'. Good quality hotels and game park lodges offer great

The development of industry and infrastructure 191

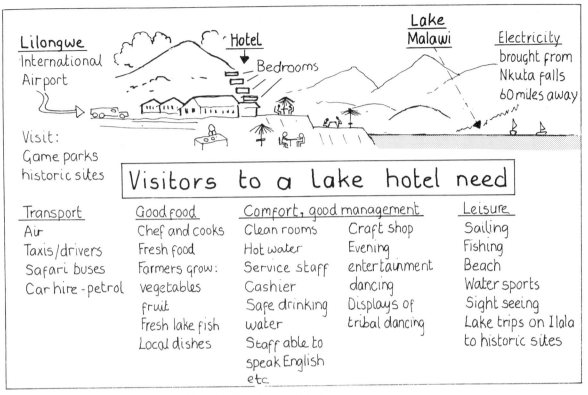

Figure 22.1 A tourist hotel by Lake Malawi: What the tourist industry needs

opportunities to tourists to enjoy game-viewing, water sports, scenery, mountain climbing and sunshine.

Many countries have established game parks in vast areas of dry savanna, mountains, and forests. Sadly, conservation has brought its own hazards:
- Too many elephants grazing bush in Tsavo East, Kenya, have uprooted and destroyed vegetation. This elephant-made desert will take years to recover.
- Game animals have also become reservoirs of trypanosomes from the tsetse fly.

The industrial development wheel, Figure 22.2

Study Figure 22.2.
1 The centre of the wheel, the hub, reminds us that all industrial production needs *all* these ingredients. However different types of industry need more of some than of others. For example, labour-intensive industry uses a lot of workers, a capital-intensive enterprise like a paper mill uses relatively fewer workers but needs a lot of money.

2 The outer rim of the wheel has 3 sections:
- raw materials and resources

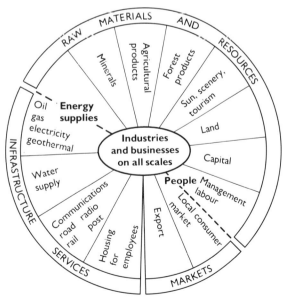

Figure 22.2 The industrial development wheel: the infrastructure support system

- infrastructure services
- markets
- Two of them overlap, because energy, that is coal, or water power, is a resource, but also related to infrastructure.
- people are both producers and consumers and so are both resources and markets.

Now look back at Figure 22.1, the tourist hotel.
1 Try to match the groups of needs in Figure 22.1 to the sections of the industrial wheel in Figure 22.2. For example, *transport* appears in the infrastructure section of the wheel. But the hotel would use all the other infrastructure services too.
2 Then go on to match up as many things as possible.

All countries hope to build a good industrial base. Small and large industries are needed to supply consumer goods (shoes, cooking pots, torch batteries); machinery for farms, harbours and railways; spare parts for all kinds of repairs (café equipment, domestic appliances, plumbing); medical supplies and so on. There are *different scales* of enterprise in industry as in most other things (see Figure 22.3). *Industrialisation* is the process of growth or change from small-scale craft-like production, to large-scale industrial centres, often clustered near each other.

But we should not under-rate the importance to every African country of the *informal sector* of industrial and commercial life (see Figure 22.4). This covers the jobs and industries that go on outside the cash records of the tax inspectors and the people who keep a country's accounts. These activities not only provide cheaper food and services but give jobs to many families and school leavers. They start businesses

Figure 22.4 A suburban pharmacy, Kinshasa, Zaire. Many pharmacies repack medicines in small quantities and translate instructions. They are good sites for informal traders

with mobile stalls and do small-scale but necessary work. Without them life would be less agreeable for those with some cash, and very hard for those without a chance to earn money.

Infrastructure: the back-up to development

Industrial and agricultural production depends on a support system of a great many things, like railways, roads, sewage works and power stations, and people who work in them. This system, called the infrastructure, is vast and complex. This section has two parts, the first is about communications; the second about power supplies. Both are key elements in infrastructure.

Communications: an example of infrastructure

Transport is often the key to successful industry and trade. Studies in other parts of the book have already shown how important road and rail routes are. Air travel has helped international communications and the tourist industry but most exports and imports depend on the land–sea system. It is not just the physical system of roads and railways that matters, it is also the reliability and speed of transfer that is important. So inefficiency and disruptions caused by wars or

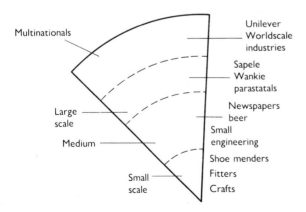

Figure 22.3 Different scales of industrial development

non-cooperation can make good systems useless.

Africa's total length of railways is not very great in comparison to other continents. But some are remarkable. The Tazara railway, built by the Chinese and opened in 1975, runs from Dar es Salaam in Tanzania to Lusaka in Zambia. A traveller or goods could continue to Bulawayo (Zimbabwe), Gaborone (Botswana) or even Cape Town (RSA), a distance of over 5,000 km.

There is still no coordinated transport network for Africa. Yet an integrated rail–road network is essential. It can only be built and run with inter-country cooperation. The route for Zambian and Zaire copper exports shows how essential this is (see page 41). If land-locked countries are to survive, the gaps must be filled and the system kept working. This is vital for large-scale exports but also for essential supplies of small-scale imports for country stores, spare parts for repairs, etc. The SADCC countries (see page 122) are working hard on this. Improved air and telecommunications links are vital.

Groups of countries are combining to plan transport systems. One of the most impressive on paper is the ECOWAS road plan (see also Chapter 9, page 89). Because of the way west Africa was divided up by colonial powers, it is easier to travel from the coast inland than from east to west. Now one of the major tasks of ECOWAS is to build or improve 10,000 km of interstate highway. Most important are the two east–west roads, one near the coast, the other parallel but roughly 800 km inland.

The coast road will link ECOWAS capitals from Nigeria in the east to Nouakchott in Mauritania. 800 km are at present only earth roads or tracks impassable in the rains. A further 900 km are gravelled. All will be bitumen surfaced, but this will take a long time (see the photograph, Figure 22.5).

The inland road will pass through the Sahel capitals from Dakar in Senegal to Ndjamena in Chad. Some sections are tarred, and 700 km gravelled, but there are still 1,300 km of earth road (see Figure 23.6).

Power resources and energy supplies

We have seen how important road and railways are. But road trucks and railway locomotives need fuel: diesel oil, coal or electricity. And extracting minerals and operating industries is impossible without energy supplies. So they are all interlinked.

Figure 22.5 Road construction through oil palm bush in Nigeria. Note how it is built up above swamp/flood level

The industrial 'revolution' in Europe and America was based on good supplies of coal. Only the southern half of the African continent has rich coal deposits. This means that most places north of Hwange (Zimbabwe) are dependent on energy based on mineral oil or hydroelectric power (HEP).

The Owen Falls dam (Uganda) was built in the 1950s to provide electricity for Uganda and especially for mining and smelting copper. This was also the reason for building the Kariba dam in the 1960s. There is enormous potential for developing electrical power from the waters of Africa's great rivers and lakes. There is more about these projects in Chapter 23.

The extraction of mineral oil in Africa also dates from the second half of this century. It has transformed the economy of oil producing countries.

Look back at one of these sections: Libya, page 106, Nigeria, pages 86–88, Algeria, page 100, Gabon, page 31.

Most countries are using oil for generating electricity as well as for fuelling cars, lorries and locomotives. But since the steep rise in the price of oil in 1973, countries which do not produce their own are trying to reduce oil imports which cost far too much. They are looking for alternative energy sources.

Two examples of this follow:
- How Kenya is developing electricity from heat below the ground; geothermal electricity.
- How appropriate technology can help, using simple equipment and readily available sources.

194 The continental view

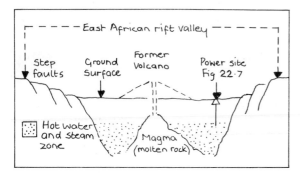

Figure 22.6 A section across the East African rift valley showing the source of geothermal heat below the rift valley floor

Geothermal power

The twenty-first century may see much greater use being made of this sort of power. Geothermal means earth heat. We all know of the immense power of volcanic eruptions. There are hot grounds in many parts of Africa (see page 16, Cameroon). Kenya is the first country in Africa to generate electrical power by using steam heated far below ground in the East African Rift Valley.

Figure 22.6 shows a section through the Rift Valley in Kenya.

Note:
1 Tension (pulling apart) resulted in step faults and allowed the centre block of the Rift Valley to sink.
2 At one time there was an active volcano, pouring out lava and ash during eruptions.
3 The materials came up the volcanic pipe from the molten rock (*magma*) zone of the earth's core.
4 The *magma* is still near to the surface in this faulted zone.
5 It heats the rock above it, and this in turn heats rain water that has sunk in. This reaches boiling point and turns into steam.

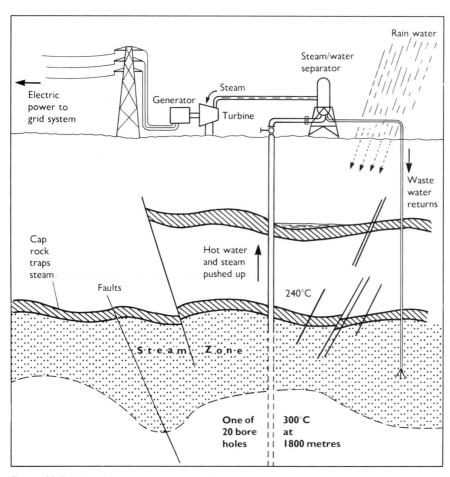

Figure 22.7 How electricity is generated from geothermal heat

Now use Figures 22.6 and 22.7 to make sure that you understand how the geothermal system works.

6 Find the power site on the Rift Valley floor.
7 Find the steam zone. (The steam is trapped by cap rock beds.)
8 If bore-holes or wells are sunk through the cap rock, the steam bursts out up to the surface. (Remember how steam can lift the lid off your kettle?)
9 It is trapped in the *separator* and the steam is diverted to the *turbine* which generates electricity.
10 Waste water is returned underground.
11 Electricity is fed into the power grid.

Many other rifted or volcanic areas could use earth heat to generate electricity.

- High rainfall is needed to recharge the underground water (aquifer). Water is also needed for drilling bore-holes, etc.
- Geothermal energy is renewable, very valuable if other mineral sources run out.
- Electricity is fed into a country's power grid in the same way as coal or water generated power.

Appropriate or intermediate technology

When people talk about power supplies they are usually thinking on a large scale. Geothermal power fits in at this level. But what often matters to ordinary people, whether they live in villages, in towns or in the bush, is what they can use to cook food and boil water every day of their lives. So fuel is a priority.

Alternative sources of fuel are needed. Experiments could be tried at the household and small workshop level. Improved cooking stoves burn slowly and save wood. Small village biogas plants could be built. Biogas is methane made from rotting rubbish and dung. The soggy residues from this can still be used as fertiliser. Some factories recycle waste products to provide power, for example, the timber mill at Sapele, and the sugar mill at Mumias. Residues from sugar processing can also be distilled into alcohol for fuel. By-product gas from coke making at Wankie is used to fuel the coke processing ovens! The best energy sources are those that are renewable, like the wind, the sun, and water, if they can be harnessed.

People also need help with everyday farm jobs like hoeing and ploughing, and reliable transport to get to work. Tractors are often unreliable, they need expensive diesel oil and skilled mech-

Figure 22.8 Small scale technology undergoing trials in Botswana. A bicycle wheel and pedals are used to grind edges on lenses for low cost spectacles

anics and spare parts when they go wrong. There are many different makes of tractor and the parts come from abroad and are often not available. Often the only way of getting parts is to dismantle one tractor and use its parts to make others work. Useless, rusting tractors are found in tractor 'cemeteries' all over Africa. Learning how to make stronger hoes and spades that do not break can be a more long-term solution than buying a tractor.

The best improvements are sometimes those that keep local practices and introduce a suitable new implement, like a plough that can be pulled by one ox rather than two as most farmers have only one ox anyway. Countries need large projects that work. They also need appropriate technology and ideas that work simply and cheaply (see Figure 22.8). Check back to:

- coal bricks described in Chapter 15, page 134
- making terraces in Burkina Faso, page 178.

Involving women in thinking up ways of making their work easier could produce surprising results.

Chapter 23 Planning a continent

> **Key words**
>
> Coordination, cooperation, research, river basin development, pollution control, gaps, social justice

At the beginning of Part 3 we said that Africa is being 'planned' in a way no other continent has been. The main coordinating organisations are the agencies of the United Nations (UN) and the Organisation of African Unity (OAU).

International planning organisations

The United Nations is the world 'umbrella' organisation concerned with development in Africa. Its Economic Commission for Africa (ECA) is sited in Addis Ababa in Ethiopia, with subregional offices in other parts of the continent. What is clear is that if we are to understand what is going on in Africa, we have to learn a new 'language': the ABC of capital letters. These are the shorthand names of dozens of projects and agencies now working in the continent. We have already met UN, ECA, and OAU, and in earlier chapters we have met WHO, FAO, UNICEF, UNESCO, GNP, SADCC and ECOWAS. One of these is not an agency or organisation. Which is it?

How many of the following do you recognise?

ACP–EEC Convention	Sometimes called the Lomé Convention (African, Caribbean, Pacific states–European Economic Community)
OPEC	Organisation of Petroleum Exporting Countries
UNCTAD	UN Conference on Trade and Development
UNDP	UN Development Programme
UNEP	UN Environmental Programme
IMF	International Monetary Fund

Coordination and cooperation, or conflict?

While it is also concerned with development, the OAU at its summit meetings is trying to steer the continent on a path that reduces friction and encourages cooperation. Sadly there is a great gap between people's hopes and what actually happens.

There are so many armed clashes in the continent that one is bound to ask why. Figure 23.1 shows some of the flash points. In the last 25 years there have been over 60 violent changes of government in African states, half of them successful. These coups are relevant to the study of the geography of the countries concerned. They disrupt the economy, cause U-turns in policy and sometimes actual changes of frontier.

The causes vary: sometimes it is inter-tribal rivalry, competition for scarce land, or belief in conflicting political systems. Sometimes one country tries to take over part of another. Is this

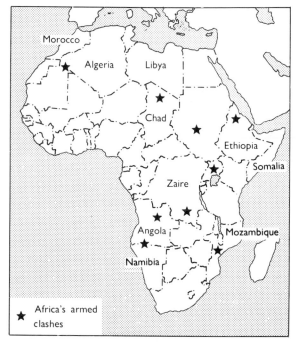

Figure 23.1 Flashpoints in Africa

part of a new wave of African imperialism: a new scramble for Africa?

Why is this happening?

The division of Africa between colonial powers in the nineteenth century left many traditional African nations divided: the Yoruba in Nigeria and Benin, the Bakongo in Angola, Zaire and Congo, the Maasai in Kenya and Tanzania, the Ewe in Ghana and Togo (see also Chapter 3, page 26). Also some very different peoples have been 'lumped together' as in the Sudan where the black Christian peoples of the south are very different from the Muslim peoples of the north. Within a country this means that there are now peoples who do not feel 'nationhood' or a sense of belonging to each other and resent being ruled by a different group.

Where differences are resolved and a strong government is created, different tribal groups and blacks and whites can work together, as in Zimbabwe and Zambia, Ivory Coast and Botswana. But where external politicians encourage the differences between rival African interests they foster disputes.

The OAU is also concerned with the plight of Africa's 3½ million refugees, half the world's total. Some estimates make this total 5 million. There are refugees in nearly 30 African countries. The shows where they are.

AFRICA'S STATELESS MILLIONS		January 1985
Host country		Refugees chiefly from
Algeria	167,000	Western Sahara
Angola	92,000	Namibia, Zaire
Burundi	256,300	Uganda, Rwanda
Cameroon	13,700	Chad
Central African Republic	42,000	Chad
Djibouti	16,700	Ethiopia
Ethiopia	59,600	Sudan
Lesotho	11,500	RSA
Rwanda	49,000	Uganda
Somalia	700,000	Ethiopia
Sudan	850,000	Ethiopia, Uganda
Tanzania	179,000	Burundi, Rwanda
Uganda	151,000	Rwanda, Zaire
Zaire	317,000	Uganda, Angola
Zambia	96,000	Zaire, Angola
Zimbabwe	46,000	Mozambique

This includes only countries where more than 10,000 refugees are present. Non-registered refugees or persons displaced within their own country are not included.

'In the country where he was born a man is at home with the smell of the soil and the feel of the wind.'

The list reflects a wide variety of situations: the Sahel drought migrations, political harassment, and too many real wars. Increased prosperity for African peoples depends on cooperation. But where a country has twenty languages even finding a common language is a problem. Most Africans are multi-lingual. One positive contribution made by the colonial period is the international languages, English and French, which now provide a means of international communication and access to books and information not yet available in Africa's own languages.

Cooperation is needed at every level, and particularly for large-scale development planning. This is quite clear when it comes to the development of water resources.

African cooperation: the need for water development

Africa has 40 per cent of the world's potential for developing water resources. The physical features of Africa make it easy:
• Rivers from the high-rainfall equatorial centre, often drop steeply over the rim of the African plateau on their way to the sea. These falls are excellent power sites.
• Warping and rifting are responsible for huge lakes; some of these act as reservoirs for water storage.
• The main uses of water developments are for irrigation, and hydroelectric power. Africa has suffered from long cycles of drought for centuries. Irrigation water is most needed in the savanna.
• Irrigation projects can have other benefits such as flood control.

- Hydroelectric power projects provide essential cheap electricity and can act as spur to industrial development.

But there are problems, for example, people often have to be moved to make way for the huge lakes; bilharzia is sometimes increased; and soils become saline when water evaporates and leaves chemicals behind.

Many major developments in Africa so far have been dams. The fact box, page 199 shows the major water projects of Africa. Figure 23.6, on page 205, shows where they are.

Earlier dams usually had only one main purpose, for example irrigation in the Gezira (Sudan), or hydroelectricity at Owen Falls (Uganda). Later dams are multi-purpose, and the 'spin-off' from all dams can be multiple. They include fisheries, transport, tourism and other aspects of everyday life. But the large river basins of Africa really need an overall development strategy, covering a long period.

The financial and planning preparations are enormous using a whole range of disciplines. The planners need:
- to see into the future and assess the benefits of different plans
- to balance the technical data from engineers, hydrologists (irrigation, dam building), agronomists, (crop systems, pests), sociologists (effect on families), politicians, etc., and advise a first choice for action.
- to present a clear picture of the likely results (the impact) and suggest alternative paths if the situation changes.

Figure 23.2 The Kariba Dam and power installations on the River Zambezi. Lake Kariba stretches for 400 km along the middle course of the river and displaced 50 000 people. Building any dam alters the ecological balance. Drowned trees later ruined fishing nets and encouraged the Kariba weed which choked shorelines.

'Look after this planet: it's the only one we've got.'

The studies that follow show:
- The Zambezi as an example of river basin development strategy.
- The need for regional cooperation and coordination: an example from the Gulf of Guinea.

The Zambezi Action Plan

The River Zambezi is a good example of the need for cooperation in river basin planning and integrated development when a great river is shared by several countries. The dam and lake at Kariba show how a large project affects the people, the landscape, and the economy.

In May 1987, 5 countries agreed to Zacplan: the Zambezi Action Plan for the joint management of the 2,240 km river. The basin with all its tributaries is vast, but its total population is only about 20 million. The tsetse fly, poor soils, and difficult climates have hindered development.

Zacplan is concerned with agriculture, mining, hydroelectricity, pollution, tourism, wildlife and water conservation.

The Zambezi rises in the far north-west of Zambia and then enters Angola. It flows south to form the border with Namibia in the Caprivi Strip. At the Victoria Falls it forms the boundary between Zambia and Zimbabwe. It flows on through Lake Kariba and into Mozambique, through Lake Cabora Bassa, to the sea. There is an immense variety of landscapes and land use. The middle Zambezi is a World Heritage site because of its unique wildlife. The Victoria Falls are one of the world's wonders, yet only a few kilometres to the south lie the huge coal deposits at Hwange. Lake Kariba produces 40,000 tonnes of fish each year. The great dams of the Zambezi basin (see Figure 23.2); depend on the flow of the river and the way it is shared. All the countries concerned will have to get the consent of the other countries in the Zacplan Accord before starting on development projects.

Facts: River basin development

River basin or river	Country	Dam or development	Purpose and details
Nile River Basin			*One of the 3 fully coordinated river systems in Africa. Integrated planning possible partly because of long tradition of irrigation, and British colonial influence in 19th century (pages 107–108).*
Nile	Egypt	• Aswan I • Aswan High Dam	1902. Irrigation and flood control. 1970. 100,000 people moved. For irrigation, flood control, and over half Egypt's electricity (pages 112–113).
Blue Nile	Sudan	• Sennar • Roseires	1925. Irrigation for Gezira (pages 116–118) 1968. Kenana irrigation (400,000 ha). Flood control and electricity.
White Nile	Sudan Sudan	• Djebel Aulia • Jonglei canal	1937. Flood control and water conservation. 240 km canal to by-pass Sudd swamp. Water conservation by reduced evaporation, increasing the flow of the White Nile (page 119).
Atbara	Sudan	• Khash el Girba	Irrigation for people moved from Wadi Halfa (Lake Nasser) (page 118).
Victoria	Uganda	• Owen Falls	1959. Hydroelectric power. Lake Victoria acts as natural reservoir. No people moved.
Niger River Basin	Mali Nigeria	• Sansanding • Kainji	1947. Irrigation and flood control (page 76). 1968. Multi-purpose: hydroelectric power, irrigation, tourism, fisheries (pages 77–78).
Zambezi River Basin			
Zambezi	Zimbabwe/ Zambia	• Kariba	1960. Hydroelectric power for both countries. Fishing, tourism and irrigation. 50,000 people moved (page 198).
	Mozambique	• Cabora Bassa	1974. Multi-purpose. RSA buys electricity. Irrigation potential.
Kafue	Zambia Zambia	• Kafue • Iteshi Teshi	1972. Hydroelectric power. 1976. Hydroelectric power.
Orange/Vaal River basins			*The 2nd fully coordinated system.*
Vaal	RSA	• Vaal	1928. Water for Witwatersrand.
Orange	RSA RSA	• H.F. Verwoerd • P.K. Le Roux	1971. Multi-purpose: water supply, hydroelectric power and irrigation 1977. Hydroelectric power and irrigation.
Tugela	RSA	• Pumping scheme	1974. Water lifted 500 m vertically to Vaal basin.
Others			
Tana	Kenya	• Sequence of dams	Multi-purpose: irrigation, hydroelectric power, water control.
Volta	Ghana	• Akosombo	1966. Hydroelectric power, fisheries and transport. 78,000 people moved.
Zaire	Zaire	• Inga 1/2	1972/77. Hydroelectric power (page 42).

Pollution in the Gulf of Guinea

Some problems cannot be solved without countries agreeing to try to improve a dangerous situation. About 20 million people live near the shores of the Gulf of Guinea. The population is expected to double in the next 25 years. Yet only one town, Tema, the chief port of Ghana, has a proper sewage system – if it is kept working properly. A complex pollution problem is building up:
- There are lagoons along much of the coast where sewage and other refuse is trapped.
- Several countries are oil-producers and there are inevitable oil spills and waste.
- Waste and water from industrial processing of sugar, timber, phosphates, alumina and iron ore pour into the sea.
- Agricultural pesticides contaminate run-off.

Three things are likely to suffer:
- health
- the tourist industry
- and fisheries

The Guinea coast has a thriving fishing industry and very good potential. 'Herring' and 'shrimp' are fished by traditional methods including beach seining (using a horseshoe shaped seine net). Powered boats fish both deep sea and coastal fish. These and the oyster beds are at risk from pollution which has already affected some lagoons, killing all marine life. UNEP is now sponsoring research and planning as part of the Gulf of Guinea Action Plan but a policy cannot be operated without joint action from all the countries.

Africa and the rest of the world: trade and aid

The problems of the trade deficit and of massive debt repayments have featured throughout the book (see Chapters 5, 12, 21 and elsewhere). The trouble is that many African countries are earning less but everything costs more.

> 'A basket of African exports buys a third less than a decade ago.'

- In order to buy what they cannot produce for themselves countries must sell goods or services.

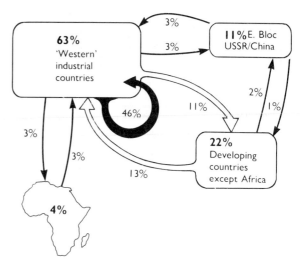

Figure 23.3 Who shares in world trade?
- The figures in the diagram are percentage shares of world exports
- Africa's share is 4%
- The arrows show the movement of *visible* exports and imports
- The "western" industrial countries include Japan, Australia and New Zealand. They dominate world trade (63%) and form a club that trades within itself (46%)
- Africa is on the edge of world trade

All African countries are competing for markets. The agricultural and mineral products which form the bulk of their exports are primary raw materials which are worth much less than manufactured goods. Often if a country tries to increase its income by growing and exporting more of something, it may cause a glut and a drop in prices because several more countries are also pouring more onto the world market. Thus a country may produce more but earn less.
- As a result of world inflation, that is, a fall in the value of money, many manufactured goods cost more. Oil has increased greatly in value which helps some countries but hinders others.

Nearly half the world's trade goes from one industrial country to another (see Figure 23.3). Africa produces only 4 per cent of the value of the world's exports. There are so many trade agreements that present links and preferences make it difficult for a country to break into new markets. African countries need new trade opportunities to stimulate home production.

When a country has a trade gap, that is, it is

in debt to other countries, it has to borrow more money to pay these debts, and then borrow more money again to pay the interest on these loans. Many African countries keep going on loans from agencies such as the World Bank and International Monetary Fund as well as other Western countries. Some of the money is lent, and some is given as 'aid'. But most aid has 'strings', or conditions, attached (usually to do with buying things from the donor country). It is now generally recognised that more money is paid back in interest on loans each year by many African countries than is given as aid by Western countries.

Aid: does it help?

Television pictures of children starving in the Sahel drought shocked the world and triggered one of the biggest ever international efforts to bring aid to famine-stricken countries.

Immediate help was essential, but tragedy like this could strike again. What is the use of sending food to save children now, if they are likely to starve later? Direct food aid is essential in emergencies but it can have the effect of lowering the market prices to local farmers who then have no incentive to grow food. The real task is put money into long-term strategies which will help Africa to feed itself. Many agencies are involved in just such schemes (see Chapter 20).

Critics of international aid argue that African countries have become too dependent upon it. This can have 2 effects:
- The money is spent on grandiose large projects which do not increase food production.
- Governments lose control over their policies: aid is often given on the condition that certain economic policies are adopted (see Chapter 12).

Peace and justice: the chief needs for development

Throughout the book we have emphasised that disputes and wars waste resources and ruin lives. Is peace the most important need for development, that is, a better life for most people? If so, the second most important need must be social justice. By social justice we mean the right of everybody to share in these better living standards.

We have mentioned trade gaps earlier in the chapter. In Chapter 12 we talked about the 'food gap' in Egypt. There are many 'gaps' that need to be reduced.

1 *The trade gap (deficit) between income and spending, imports and exports.* If it seems impossible for countries to pay off their huge debts they will not struggle to live within their incomes. If these debts are written off African countries must try to do their part.

2 *The gap between rising population and food supply.* A country can only afford more and more people if it increases food production or makes enough money with exports to import food. Both family limitation and the long-term development of resources are needed.

3 *The gap between the 'haves' and 'have nots'.* Sometimes a small group gets rich at the expense of the rest. The social and economic gaps are increasing in many African countries. It is normal for people to want to do the best for their children and to 'get on'. How can people find ways of reconciling their own interests with the well-being of others? How can a country reward the people with the ability and technical know-how essential to its development without creating an elite? Social justice means that all people must all have a right to share in a country's better living standards.

4 *The gap between men and women in status and rewards for hard work.* Women need involvement in making decisions, access to training, and a share in the incentives.

The table on page 202 gives a checklist, a summary of thoughts, ideas, and hopes for the future.

'What can *people* do?
- support the family
- talk about what they want
- take the responsibility for carrying things out
- find incentives for farmers and women
- keep developments small: "small is beautiful"
- work together on things they *want* to do
- look after the land: the final resource.'

A checklist for the future

Growth

• People and countries must earn money to buy things they cannot produce themselves.	Balance of payments
• If people and countries earn cash they can both spend and save.	'Take off'
• Spending feeds money into the market and stimulates still more jobs and industries.	'Multiplier effect'
• Savings can be invested in businesses and infrastructure.	'Up-spirals'
• If people and countries produce some of the food and goods they need, they do not have to buy expensive imports.	Import substitution
• A flourishing home market encourages small businesses.	Build home markets

Regional variations

• Some places have advantages of location or resources. They are the 'winners'. Others, less well endowed, are 'losers'.	'Winners or losers'
• Enterprising people from less favoured areas often move to areas of opportunity in search of a better chance of getting a job.	Growth areas
• This can increase the gap between the favoured and less favoured areas.	Reduce gaps
• Governments can take action to help the disadvantaged areas and reduce the gaps. They can give support to enterprising local people.	Local planning

So what might be the strategy for planners?

• Find ways of combining the small and large scale, and African as well as Western, life-styles. Foster the traditional informal sectors as well as large projects.	Dualism
• Learn to live with change; keep the good in the old ways.	Flexibility
• Small scale is less likely to be seriously wrong and more likely to fit in easily. When large projects fail the results are devastating, but small failures can be absorbed.	Appropriate technology
• Ask the local people what they need.	'Grass roots'
• Keep a reasonable balance between town and country.	Rural development
• A country can feel its way slowly towards improvement and growth.	'Evolution' not 'revolution'

At the international level

• Avoid costly disputes and overspending on defence.	Avoid military spending
• Avoid loans that have to be repaid, plus interest.	Avoid debts
• Use UN agencies that can offer experience, research and money.	Use aid
• Accept all possible help from research projects related to agriculture, health, etc.	Promote research
• Use satellite scans to help with weather forecasting, soil analysis, crop planning.	Use international technology
• Countries can usually achieve more if they cooperate. (The OAU could do more if it had cash support.)	Work together
• Act as guardians of the earth's resources.	Conservation

Planning a continent 203

Figure 23.4 Hope for the land. Trees are planted in Somalia to retain soil moisture and fences built to protect crops from grazing animals

The photograph on the back cover shows part of the central area of Harare, Zimbabwe. Find the line drawing of this photograph on page 144.

Facts: Different types of examination questions

1 Objective questions

These are called objective because there is no doubt about whether the answer is right or wrong. There are different types of objective questions but in all of them a choice of answers is provided.

- The most usual type is called *multiple choice*. Students are given up to five choices, only one of which is correct. For example:

Irrigation is: (Tick √ the correct answer)

1	Draining water from a flooded area
2	Using wet land for crops after a flood
3	Watering crops in dry seasons
4	Storing water in a reservoir

- A *multiple completion* question is similar, but there can be more than one correct answer.
- A question which asks students to *match* 'heads and tails' of information correctly, usually has between 2 to 4 items. The example on page 175 is longer than this.

2 Resource based questions

These are now common in geography. They can be tackled successfully with care and a little knowledge. The information is there in front of you, and you should be able to draw conclusions from it without having learned all the facts. Look back at practical work box number 7, on page 104.

3 Written examination questions

These are the most difficult, because you have to:
- decide what the question means and what the examiner wants
- think carefully about what you know about the subject
- select, organise and present your knowledge in a short, clear, way

These are not 'write all you know' questions, and you will not get high marks if you do this. What matters is how well you organise your facts.

204 The continental view

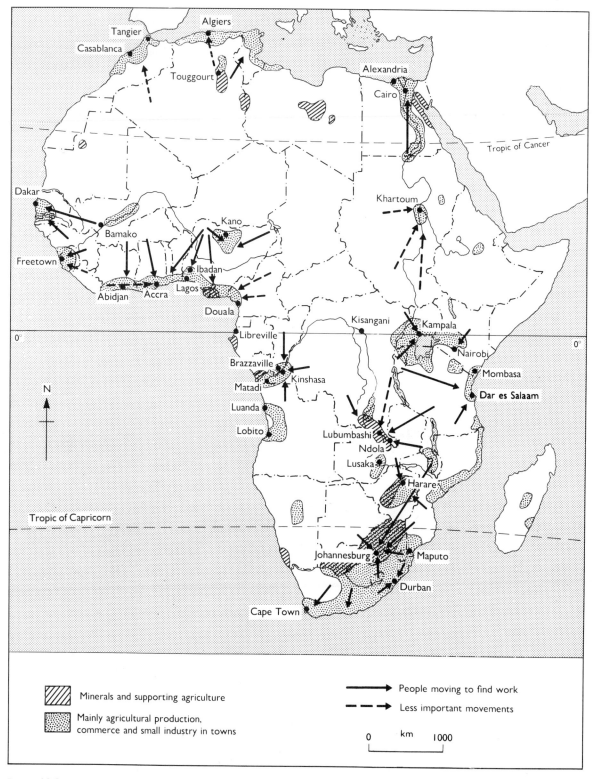

Figure 23.5 Areas of opportunity in Africa

Planning a continent 205

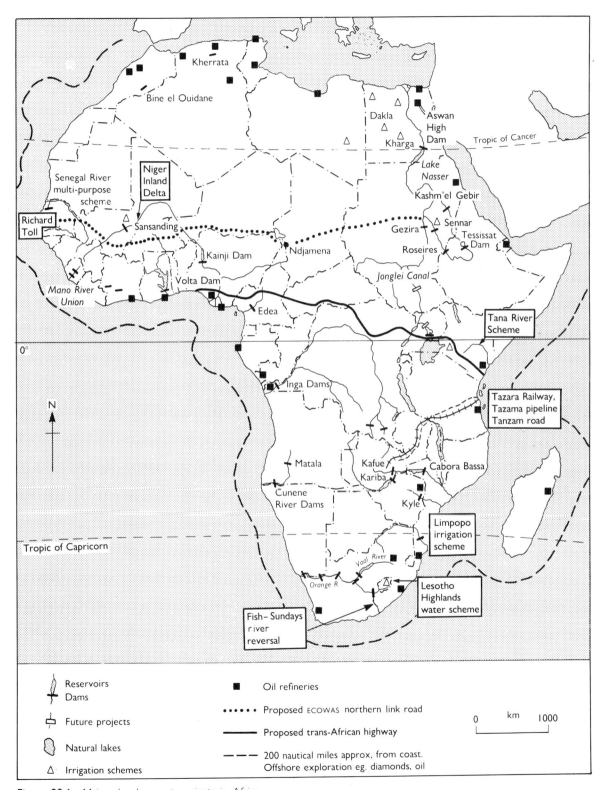

Figure 23.6 Major development projects in Africa

Index

Page references in italic refer to illustrative material, in bold to trade summary tables.

A
Abidjan, 82, 184
Abuja, 85–86, 188
Accra, 20, 82
acquifer, 102, 108, 195
agriculture – themes table, 12
agro-forestry, 179
aid, 200–201
AIDS, (Acquired Immune Deficiency Syndrome) 171
Akwapim, 68–69
Algeria, 94, 100, 106
aluminium/alumina, 71
Angola, **125**
Apapa, 85, 96
apartheid, 153–155, 162
appropriate technology, 195
Archimedean screw, 111
Aswan High dam, 107, 112–113
Atlas mts, 91

B
bananas; fact box 182
basement complex, 14, 37–38, 121, 122, 146
bauxite, 71
Bawku, 79
Beira corridor, 122–123, 144
Benguela railway, 41, 122, 125
bilharzia, 171, 198; fact box 172
biogas, 195
birth rate, 169
Black homelands, 151, 153–154, 162
Blida, 94, 94–95
Bophuthatswana, 155
Botswana, **124**, 125, 126, 170
Burkina Faso, 79, **80**, **81**, 178
Burundi, **60**

C
Cabora Bassa dam, 125
Cairo, 114–115, 184
Cameroon, 71
Cape Town, 152
capital cities, 85–86, 187–188
case studies, 7
cassava; fact box 182
cattle, 124, 135–136, 140
Central African Republic, **35**
central business district (CBD), 58, 84, 85
Chad, **80**, **81**
charcoal, 124
chitemene, 134
climate: 16–20, 21, 23
    East Africa, 45–46
    figures, 24
    West Africa, 61–62

coal, 128, 134, 140–143, 150, 151; fact box 141
cobalt, 40, 41, 121
cocoa, 68–70, **71**; fact box 70
coffee; fact box 49
colonial boundaries, 26, 27, 197
Commonwealth Development Corporation, 85
communal areas (Zimbabwe), 138
communications – see transport
compound homestead, 72, 73, 74
Congo Republic, **35**
consolidation (of land holdings), 111
copper, 40, 41, 127–129, 130–131
Copperbelt, 41, 127–129
cotton, 116–118; fact box 120

D
Dakar, 82, 184
dams: 205; fact box 199
    Aswan, 107, 112–113
    Kainji, 75, 77–78
    Kariba, 193, 198
death rate, 169
deserts, 101–106
desertification, 106, 160, 177–178
diamonds, 40, 41, 121, 124, 125, 126
disease, 170–173; fact box 172
diversification, 50, 134
Dodoma, 188
drought, 78–79, 106, 120, 176–177
dual economy, 138
Duka township, 56
dust storms, 106

E
eastern Africa, 43–60
Economic Commission for Africa (ECA), 196
Economic Community of West African States (ECOWAS), 85, 88–89, 193
economic development, 165–167
ecosystems, 23
Egypt, 107–115, **112**
Eritrea, 119
Ethiopia, 116–118, **119**
ethnic groups, 27
evapotranspiration, 19
examination questions, 104; fact box 203

F
famine, 79, 176
farming – themes table, 12
firewood, 177, 179–180
fish farming, 180

fishing, 76, 77, 78, 81, 113, 180, 200
flow diagrams, 10, 64, 129
fold mountains, 14
food, 176, 178–179, 181
forests, 31–32, 34, 62–67, 69
fossil water, 102, 106
Front-line States, 121–124
functional areas, 56, 58, 83, 149

G
Gabon, 31–34, **34**
Gaborone, 123, 124, 184, 188
Gambia, **81**, 174
gas, natural, 86, 87, 100, 105
geothermal power, 60, 194–195
Gezira, 116–118, 120
Ghana, 68–70, 72–74, **90**
glaciated landscape, 98–99
gold mining, 40, 121, 146–148
Great Dyke, 133, 137, 140
Great Rift Valley, 38
gross national product (GNP), 29, 34, 36, 145, 146, 165–166
groundnuts, 81
Guinea, **71**

H
Harare, 144
Harmattan, 61
history, 25–27; fact box 26
Hwange, 140–143
hydroelectric power (HEP), 193, 197, 198

I
income, national, 166–167
industrial development, 189–190
informal sector, 83, 181, 186, 192, 193
infrastructure, 86, 189–190, 191, 192–193
intermediate technology, 174, 195
International Labour Organisation (ILO), 85
Inter-tropical Convergence Zone (ITCZ), 17
iron ore, 31, 71, 106
irrigation, 59, 104, 108–110, 111, 114, 116, 117, 118, 120, 157, 160–161, 197
Islam, 26, 91
Ivory Coast, **71**

J
Johannesburg, 149–150
Jonglei canal, 119

K

Kafue river, 127
Kainji dam, 75, 77–78
Kalahari desert, 124
Kano, 82
Kariba, 193, 198, *198*
Kenya, 51–54, 59, *59*, *126*
Kenya, Mt (map extract), 98–99
Kigezi, 44
Kikuyu, 51
Kilimanjaro, Mt, 45, *45*
Kinshasa, *13*, *192*
Kitwe – see Nkana-Kitwe
kwashiorkor, 171, 181

L

labour supply, 147, 190
lagoon coasts, 96
Lagos, 82, 83–85, 86, map extract, 96–97, 184
Land Apportionment Act, 137
land development, *51*
landlocked states, 7, 121, *123*, 136, 144, 193
landscapes, 14–23
landscape sketches, *48*, *144*, *159*
Lesotho, 161, **163**
Liberia, 63, **71**
Libya, **106**
Lilongwe, 124, 188
line drawings, *10*, *33*, *48*
locusts, 171; fact box 172
logging, 32–33
Loursnsford, 157–158
Lusaka, 187, 188

M

Maasai, 46–47, *47*
Mafikeng, 188
Maghreb, 92–95
Makwiro (map extract), 132–133
maize, 156–157; fact box 182
malaria, 171, 172
Malawi, **124**
Malawi, Lake, *191*
Mali, 75, 76, **80**, **81**
malnutrition, 171
Matopos Hills, *139*
Mauritania, 106, **106**
micro-climate, 66
migration, *25*
millet; fact box 182
minerals:
    copper, 127–131
    Northern Africa, *105*
    oil – see oil
    South Africa, 145–148, 151
    Southern Africa, *121*, 122
    West Africa, *90*
    Zaire, 38, 40–42
    Zambia, *135*
    Zimbabwe, *137*, 140
Mombasa, 56, 58–59, *58*
Mopti, 76–77, *77*
Morocco, 95, **100**, *102*
mosquitoes; fact box 172

Mozambique, **125**
Mukono, 56–57, *56*
multiple-cropping, 67, *104*
multinational companies, 39
Mumias, 52–55, *53*, *54*

N

Nairobi, 185–186
Namibia, **126**
Nasser, Lake, 107, 112–113
natural gas – see gas
Niger (Republic), 80, **81**
Niger Inland Delta, 75, 76–77
Niger, river, 37, 72, 75–78, 199
Nigeria, 64–66, 67, 83–88, **90**
Nile delta, *112*, *115*
Nile, river, 37, 107–114, 199
Nkana-Kitwe, 127–129, map extract, 130–131
Nouakchott, 106, 188

O

oases, 102–104, *105*, 113–114
Ogaden, 119
oil, mineral, 31, 86, 87, 105–106, 193
oil palm, 38–39, *39*, 67–68
Okavango swamp, 124
opencast mining, 140–143
Organisation of African Unity (OAU), 164, 196, 197
outgrowers, 53
Owendo, *33*

P

palm oil, 38
palm products; fact box 38
pears, 157
pests; fact box 172
petrochemicals, 100
petroleum – see oil, mineral
phosphates, 81, 100
physical geography checklist, 23
physical landscapes, *22*
pie chart, 165
plantation agriculture, 39
plateaus, 15, *15*, 43–44
pollution, 200
population, 27–30, 168–170
    census, 168
    clusters, 28
    density, 27, 36, 60, 102, 124
    distribution, 27, *28*
    figures, 29
    increase, 169, 184, 201
    migration, *25*, 25
    pyramids, *138*, 168–169, 170
ports, *56–59*, *85*, *97*, 152–153
power resources, 86–87, 190, 193–195
    coal, 140–143
    geothermal, 60, 194–195
    hydroelectric, 40, 60, 78, *95*, 112, 125, 129
Pretoria–Witwatersrand–Vereeniging (PWV) triangle, 150–152, 161

R

railways, 51, 122–123, 144, *150*, 193
rain forests, 32, *32*
rainfall, 17–20, *17*, *21*, 45–46, *46*
raw materials, 189
refugees, 106, 117, 119, 173, 197
regional groupings, 6, 30, *30*
relief, *14*, 23
resettlement, 51–52
Rhodes, Cecil, 121
rice, 75, 76–77, *174*; fact box 182
rift valleys, 15, 44, 51, 194
river basin development, 198–199; fact box 199
river blindness, 78, 171
rubber, 63–64, *63*, 71; fact box 64
rural life – see themes table, 12
Rwanda, 60

S

Southern African Development Coordination Conference (SADCC), 121, 122, 143, 193
Sahara desert, 101–106
Sahel, 72, 78, 176, 179
Sansanding barrage, 76
Sapele, 64–66, *65*, 66
savanna, *13*, 61, 72–81, 176
Savornin's Sea, 102
self help, *178*, 185, 186–187, 188
Senegal, **81**
service centres, 54–57, 185
Shaba (Katanga), 41
Sierra Leone, **71**
simulation, 55
sisal, fact box 52
slaves, 26, 71
soil erosion, 177, *178*
soil fertility, 66–67
Somalia, *116*, *119*, *178*, *203*
sorghum (millet), 73; fact box 182
South Africa, **145**, 145–163
Soweto, 149, *150*
staple foods; fact box 182
stock rearing, 179
subsistence agriculture, 166
Sudan, 116–120, **119**
Sudd swamps, 119
Suez Canal, 112, *115*
sugar cane, 52–54, *53*, *54*, 158–160; fact box 54
Swaziland, 161, **163**

T

Tana, Lake, 107
Tanzania, **60**
tea, *50*; fact box 50
Tell, the, 91, 94
Tema, 200
terraces, 44
themes table, 12
timber production, 32, 35, 39–40, 64–66, *65*, 66
tobacco, 139–140; fact box 140
Tongaat, *158*

Touggourt, 102–104
tourism, 42, 95, 112, 113, 124, 190–191
towns – see themes table, 12
trade:
    balance, 34, 36
    deficit, 36
    gap, 34, 201
    summaries, 34–36
    surplus, 34, 36
    world, 200
transects, 48, *48*, *62*, 93
transport:
    eastern Africa, 43
    infrastructure, 192–193
    Southern Africa, 122–123
    West Africa, 89
    Zaire, 40, 42
    Zambia, *135*
    Zimbabwe, *137*
tsetse fly, 46, 73, 134, 171, 179; fact box 172
Tunisia, 95, **100**

U
Uganda, 47–49, **59**
ujamaa, 60

United Nations Agencies, 85, 171, 196
United Nations Environmental Programme (UNEP), 196, 200
uranium, 31, 81, 121, 126
urban hierarchy, *184*
urbanisation – themes table, 12 and urban geography index, 183; 183–188

V
vegetation zones, 22, *61*
volcanic areas, 15, *15*, *16*, 44, 60, 194
Volta River and dam, 199

W
Wankie colliery, 128, 134, 140–143; fact box 141
war, 119, 120, 170, *196*
Warri, *21*, 61
water:
    balance, 18–19, *19*
    resources, 79, *80*, 95, 124, *126*, 160–161, 197
    supply, *20*, 74, 77
    table, 47
wealth, 146
Webuye, 52, *59*
Western Sahara, 173
wheat, 181; fact box 182
White Highlands, 51
windpumps, 47
winds, *17*, 18, *18*
Witwatersrand, 121, 146, 149, 150
women, 173–174
World Health Organisation (WHO), 85, 171

Y
Yams; fact box 182

Z
Zaire, 37–42, *37*, *40*, **40**
Zaire, River, 37, 42
Zambezi, River, 161, 198, 199
Zambia, **127**, 127–136
Zaria, 82, *83*
Zimbabwe, 132–133, 137–144, **139**

British Library Cataloging in Publication Data
Hickman, Gladys
  The New Africa. — 4th ed
  1. Africa. Geographical features
  I. Title
  916

ISBN 0 340 39943 0

First published 1973
Second edition 1976
Third edition 1980
Fourth edition Copyright © 1990 Gladys Hickman

All rights reserved. No part of this publication may be reproduced or transmitted in any form or by any means, electronic or mechanical, including photocopy, recording, or any information storage and retrieval system, without permission in writing from the publisher or under licence from the Copyright Licensing Agency Limited. Further details of such licences (for reprographic reproduction) may be obtained from the Copyright Licensing Agency Limited, of 33–34 Alfred Place, London WC1E 7DP

Typeset by Photo·Graphics, Honiton, Devon.

Printed in Great Britain for Hodder and Stoughton Educational, a division of Hodder and Stoughton Ltd, Mill Road, Dunton Green, Sevenoaks, Kent by Thomson Litho Ltd, East Kilbride.